T0269372

Astronomers' Universe

More information about this series at
http://www.springer.com/series/6960

John Wilkinson

The Solar System
in Close-Up

 Springer

John Wilkinson
Castlemaine, Victoria
Australia

ISSN 1614-659X   ISSN 2197-6651   (electronic)
Astronomers' Universe
ISBN 978-3-319-27627-4   ISBN 978-3-319-27629-8   (eBook)
DOI 10.1007/978-3-319-27629-8

Library of Congress Control Number: 2016934072

Cover illustration: Artistic impression of the New Horizon's probe approaching Pluto in
July 2015.
Image credit: NASA/JHU APL/SwRI/Steve Gribben.

Printed on acid-free paper

This Springer imprint is published by Springer Nature
The registered company is Springer International Publishing AG Switzerland

Preface

The overwhelming importance of the solar system lies in the fact that we are part of it; its origin and evolution are part of our own history. Astronomers have traditionally observed the solar system for the past few centuries via optical telescopes from the Earth's surface. Then in 1957, a new method of exploration began with the launch of the first artificial satellite—this event marked the beginning of the Space Age. Since this time, humans have improved the technology of their spacecraft to the point where they can now send probes deep into the solar system to places never seen before. In the past few decades, there have been many space probes sent to explore the crater-strewn surface of Mercury and the roasting hot surface of Venus. In 1969, the first humans walked on the surface of the Moon. Since then, we have placed several robotic probes on the surface of Mars and used them to search for life on this planet. The giant planets Jupiter and Saturn together with their many moons and ring systems have also undergone extensive up-close exploration by space probes such as Voyager and Cassini. Saturn's rings are arguably the most spectacular structure in the solar system, and if placed from end to end, they would reach from Earth to the Moon. The cold icy planets of Uranus and Neptune have thin ring systems and more moons than previously thought.

In 2011–2012, the Dawn spacecraft explored the asteroid Vesta before moving on to the largest asteroid Ceres in 2015. In 2014, another spacecraft called Rosetta landed a probe on the surface of a comet—a momentous occasion. And in 2015, the New Horizons spacecraft visited Pluto and provided a wealth of new information about this dwarf planet and its system of five moons.

During the past decade, astronomers have used the Hubble Space Telescope to discover other planet-like bodies orbiting beyond Neptune and Pluto, in far-out regions of the solar system called the Kuiper belt and Oort cloud. These new discoveries have provided astronomers with new insights into the origins of the solar system.

These new explorations have revealed that Earth's planetary neighbours are fascinating worlds. Today, we stand on the threshold of the next phase of planetary exploration. Many new missions are currently under way and many more are being planned.

This book explores recent advances in our understanding of the solar system, in particular the effect on this understanding that the most recent spacecraft missions and the Hubble Space Telescope have provided. This book is, therefore, a record of the many discoveries made about the solar system in recent years using the context of space technology.

John Wilkinson

Acknowledgements

The author and publisher are grateful to the following for the use of photographs in this publication.

National Aeronautics and Space Administration (NASA),
European Space Agency (ESA),
Hubble Space Telescope (HST),
European Southern Observatory (ESO),
Keck Observatory,
John Wilkinson (author).

While every care has been taken to trace and acknowledge copyright, the author apologises in advance for any accidental infringement where copyright has proved untraceable. He will be pleased to come to a suitable arrangement with the rightful owner in each case.

Notes: The websites used in this book were correct at the time of writing.

Contents

1. The New Solar System

Highlights

- Latest definition of what constitutes a planet and dwarf planet.
- Mathematics can be used to distinguish between a planet and dwarf planet.
- Hubble discovers the first proto-planetary discs around young stars.
- The Modern Laplacian theory has been successful at making key predictions about the physical and chemical structure of the solar system.
- The Nice model and the Grand Tack hypothesis provide new ideas about the evolution of the solar system.

Introduction

For thousands of years, the movement of the stars and planets across the night sky has fascinated humans. Humans have wondered what these objects are made of, how they move across the sky, and whether these worlds contain other living beings like us.

In ancient times people noted the position of the Sun in the various seasons and its effect on crop growth. They also knew how the Moon affected the tides. And they observed objects called planets moving against a background of stars. The Babylonians even developed a calendar based on the movement of the planets visible to the unaided eye. In fact, the names of the days of our week originate from the Sun, Moon, Mercury, Venus, Mars, Jupiter, and Saturn. These objects are the classical objects of our night sky.

J. Wilkinson, *The Solar System in Close-Up*, Astronomers' Universe,
DOI 10.1007/978-3-319-27629-8_1,
© Springer International Publishing Switzerland 2016

Fig. 1.1 Stars develop everywhere we look in space. In our region of the universe they form mostly in the arms of spiral galaxies. The solar system we live in is part of the Milky Way Galaxy. The Milky Way has a spiral structure like that shown in this photograph of M83 taken by the Hubble Space Telescope (Photo: NASA/HST).

The word 'planet' comes from the Greek word meaning 'wanderers'. The Greeks observed that the planets wandered against a background of stars that remained relatively fixed in relation to each other. The band across the sky through which the planets moved was called the **zodiac**. The star groups or constellations that form the zodiac were given names of animals, for example, the constellation Leo resembled a lion, and Taurus resembled a bull.

Early Western and Arab civilisations and the ancient Greeks believed that the Earth was at the centre of the universe with the Sun, Moon and the then known planets orbiting around it. This view was challenged by Polish astronomer Nicolaus Copernicus in the sixteenth century when he suggested that all the planets, including the Earth, orbited the Sun in near circular orbits. By using a Sun-centered model, Copernicus was able to determine which planets were closer to the Sun than the Earth and which were further away. Because Mercury and Venus were always close to the Sun, Copernicus concluded that their orbits must lie inside that of the Earth. The other planets known at that time, Mars, Jupiter and Saturn, were often seen high in the night sky, far away from the Sun, so Copernicus concluded that their orbits must lie outside the Earth's orbit.

It was not until early in the seventeenth century that the German, Johannes Kepler showed that the orbits of the planets around the Sun were elliptical, rather than circular. Kepler also showed that a planet moved faster when closer to the Sun and slower when further from the Sun, and he developed a mathematical relationship between the planet's distance from the Sun and the length of time it takes to orbit the Sun once. These three proven observations became known as **Kepler's Laws** of planetary motion.

With the invention of the telescope in 1608 the Italian, Galileo Galilei, was able to gather data to support Copernicus's model for the Sun and planets. Galileo discovered four moon's orbiting the planet Jupiter; he also observed sunspots moving across the surface of the Sun and craters on the Moon. Galileo's discovery that the planet Venus had phases just like Earth's Moon confirmed that Venus orbited the Sun closer than Earth and provided support for Copernicus's sun-centered model.

One major problem restricting the full acceptance of Kepler's and Galileo's theories was that it was not known what kept the planets in orbit. People did not know how planets, once they started orbiting the Sun, could keep moving. Isaac Newton put the explanation of this motion forward in the seventeenth century. Newton put forward the idea that the Sun must be exerting a force on the planets to keep them in orbit. This force was called **gravity** and it exists between any two masses (such as a planet and a star like our Sun). Using his law of gravity, Newton was able to prove the validity of Kepler's three laws of planetary motion. Newton also showed that other types of orbits around the Sun were also possible. For example, the orbits could also be parabolas or hyperbolas. Newton also developed a **Universal Law of Gravitation**, which states:

Two bodies attract each other with a force that is directly proportional to the product of their masses and inversely proportional to the square of the distance between them.

This law means that the more mass a planet or star has, the greater its gravitational pull. This pull decreases with increasing distance from the object.

Discovering New Planets

Towards the end of the eighteenth century, only six planets were known—Mercury, Venus, Earth, Mars, Jupiter, and Saturn. In 1781, British astronomer, William Herschel accidentally discovered the seventh planet Uranus. In 1846, Urbain Leverrier in France, and John Adams in England used Newton's gravitational laws to independently predicted that variations in the orbit of Uranus were due to the influence of an eighth planet. Soon after, the Berlin observatory found the predicted planet and named it Neptune. In the early twentieth century Percival Lowell and William Pickering predicted that another planet should exist beyond Neptune. In 1930, Clyde Tombaugh found a body, which was named Pluto, close to where Lowell and Pickering predicted it to be. Between 1930 and 2006 Pluto was regarded as the ninth planet of the solar system. However, in 2006 a meeting of the International Astronomical Union (IAU) decided on a definition of a planet that excluded Pluto as a planet, making it a 'dwarf planet' along with a number of other newly discovered bodies. As a result we now have what many call, 'the new solar system'.

What Is a Planet?

Traditionally, a **planet** has been regarded as a spherical body that orbits a star and is visible because it reflects light from the star. The spherical shape is only possible when the object has enough mass that gravity is able to pull it into a spherical shape. All planets, and many large moons and large asteroids are spherical.

Fig. 1.2 Astronomers use optical telescopes to explore the universe. Pictured is the 3.9 m diameter Australian telescope on Siding Springs Mountain, Australia (Credit: Australian Astronomical Observatory/David Malin).

In August 2006 the IAU decided on the following definition of a planet:

To be a planet a body must

1. be in orbit around the Sun,
2. have sufficient mass for self-gravity to overcome rigid body forces so that it assumes a hydrostatic equilibrium (nearly spherical) shape, and
3. have cleared the neighborhood around its orbit.

What made this definition suddenly critical was the discovery of a number of objects in the outer solar system beyond Pluto. Before the definition was accepted, there could have been as many as 50 planets orbiting the Sun.

The IAU also introduced a new classification—that of a 'dwarf planet'. A **dwarf planet** is a body that orbits the Sun, has sufficient

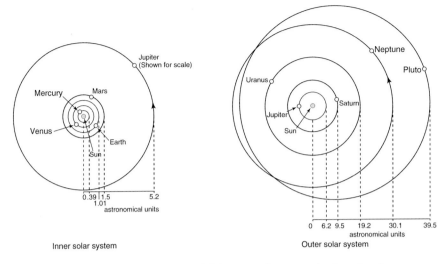

Fig. 1.3 (**a**) The inner planets and (**b**) outer planets orbiting the Sun.

mass for self-gravity to have pulled it into a spherical shape, and has NOT cleared the neighborhood around its orbit, and is NOT a satellite. All other objects orbiting the Sun are collectively known as 'small solar system bodies'. Currently, there are a number of bodies regarded as dwarf planets and more are expected to be added to the list over the next few years when more information about them is known. Examples of dwarf planets include, Pluto, Eris and the large asteroid Ceres. Dwarf planets are not considered to be true planets mainly because they do not have the ability to clear their orbital path of other material.

Currently there are eight major planets in the solar system. In order of distance from the Sun they are—Mercury, Venus, Earth, Mars (the inner planets), Jupiter, Saturn, Uranus and Neptune (the outer planets). There are also five dwarf planets—Ceres, Pluto, Haumea, Makemake and Eris; but more are likely to be added to this list as details of them are better determined (Fig. 1.3).

Many of the planets in the solar system have natural satellites or moons orbiting them. The planets Jupiter and Saturn have the most moons orbiting them. True moons are large enough for gravity to have pulled their mass into a spherical shape, while smaller moons do not have enough mass and gravity and are irregular in shape. To be a moon, a body must be naturally orbiting a planet and be smaller than the planet. The planets have captured most of their moons during the formation of the solar system.

The four largest planets (Jupiter, Saturn, Uranus, Neptune) are also surrounded by planetary rings of varying size and complexity. These rings are composed primarily of dust or particulate matter. Saturn has the most prominent rings and these can be easily seen through small telescopes on Earth. The origin of such rings is not known, but they may be left over debris from moons which have been torn apart by tidal forces (see Table 1.1).

Difference Between a Planet and Dwarf Planet

Dwarf planets are not just smaller than a planet. The main difference is that a dwarf planet does not have the ability to clear its orbital region of other matter, while a planet has. Planets are able to remove smaller bodies near their orbits by flinging them away, sweeping them up or by holding them in stable orbits, whereas dwarf planets lack the mass to do so. To make this clearer some planetary scientists have developed mathematical methods that help us to distinguish between a planet and a dwarf planet.

American planetary scientists Alan Stern and Harold Levison introduced a parameter Λ (lambda) to express the likelihood of a body "clearing the neighborhood around its orbit". The value of this parameter is proportional to the square of the mass M (which determines the gravitational reach of the massive body for a given amount of deflection) and inversely proportional to the time to orbit the Sun T (which governs the rate at which the encounters occur). If $\Lambda > 1$ then the body will have already or will eventually clear its orbital path of other matter. If $\Lambda < 1$ then the body is unlikely to clear its orbital path and would therefore classed as a dwarf planet.

$$\Lambda = k \ M^2 \ T^{-3/2}$$

where k = 0.0043.

Steven Soter (another USA planetary scientist) and other astronomers developed another method for distinguishing

Table 1.1 Major planets and dwarf planets in the solar system

Name	Class (major or dwarf planet)	Average distance from Sun (AU)	Diameter (km)	Number of moons	Ring system
Mercury	major	0.38	4880	0	none
Venus	major	0.72	12,104	0	none
Earth	major	1.00	12,756	1	none
Mars	major	1.52	6,794	2	none
Ceres	dwarf	2.76	933	0	none
Jupiter	major	5.20	142,984	67	faint
Saturn	major	9.54	120,536	62	prominent
Uranus	major	19.2	51,118	27	faint
Neptune	major	30.1	49,532	14	faint
Pluto	dwarf	39.6	2,370	5	none
Haumea	dwarf	43.1	1240	2	none
Makemake	dwarf	45.8	1430	0	none
Eris	dwarf	67.7	2326	1	none

Note: Distances are given in astronomical units (AU) where one AU is the average distance between the Earth and Sun between planets and dwarf planets also based on their ability to clear the neighborhood of their orbital path. Soter developed a parameter μ (mu) called the planetary discriminant, as a measure of the actual degree of cleanliness of the orbital region.

$$\mu = M/m$$

where M = mass of the candidate body, m = total mass of the other objects that share its orbital region.

If μ > 100 then the body has cleared its orbital path and is a planet. If μ < 100 then the body cannot clear its orbital path and would be classed as a dwarf planet.

There are several other schemes that try to differentiate between planets and dwarf planet.

Table 1.2 shows the value of Λ and μ for each planet and dwarf planet. It is clear that dwarf planets are different to planets as they do not have the ability to clear their orbital region of other matter (i.e. they have very low values of Λ and μ).

The IAU has not specified the upper and lower size and mass limits of dwarf planets. The size and mass at which an object attains a hydrostatic equilibrium shape depends on its composition and thermal history.

Table 1.2 Distinguishing between planets and dwarf planets (*)

Body	Mass (Earth=1)	Λ	μ
Mercury	0.055	1.95×10^3	9.1×10^4
Venus	0.815	1.66×10^5	1.35×10^6
Earth	1	1.53×10^5	1.7×10^6
Mars	0.107	9.42×10^2	1.8×10^5
Ceres *	0.00015	8.32×10^{-4}	0.33
Jupiter	317.7	1.30×10^9	6.25×10^5
Saturn	95.2	4.68×10^7	1.9×10^5
Uranus	14.5	3.85×10^5	2.9×10^4
Neptune	17.1	2.73×10^5	2.4×10^4
Pluto *	0.0022	2.95×10^{-3}	0.077
Haumea *	0.00067	2.68×10^{-4}	0.02
Makemake *	0.00067	2.22×10^{-4}	0.02
Eris *	0.0028	2.13×10^{-3}	0.10

As of 2015 the IAU recognizes five bodies as dwarf planets: Ceres, Pluto, Haumea, Makemake, and Eris. Ceres and Pluto are known to be dwarf planets through direct observation. Eris is generally accepted as a dwarf planet because it is more massive than Pluto, whereas Haumea and Makemake qualified to be assigned names as dwarf planets based on their absolute magnitudes. There are other bodies in the outer solar system that some astronomers like Mike Brown (USA), believe are also dwarf planets, for example, Orcus, Salacia, Quaoar, and Sedna. No space probes have ever visited these bodies.

Moons and Dwarf Planets

Many of the planets and dwarf planets in the solar system have smaller bodies called "moons" orbiting them. For example, Earth has one moon; Mars has two moons, while Jupiter has at least 63 moons. To be a moon, an object must be in orbit around a planet or dwarf planet. True moons have been pulled into a near spherical shape by self-gravity. Small moons that are irregular in shape are often captured asteroids and are better referred to as 'moonlets'.

Nineteen moons are known to be massive enough to have relaxed into a near spherical shape under their own gravity, and seven of them are more massive than either Eris or Pluto. Moons are not physically distinct from the dwarf planets, but are not members of that class because they do not directly orbit the Sun. The seven that are more massive than Eris are: Earth's Moon, the four Galilean moons of Jupiter (Io, Europa, Ganymede, and Callisto), one moon of Saturn (Titan), and one moon of Neptune (Triton). The others are six moons of Saturn (Mimas, Enceladus, Tethys, Dione, Rhea, and Iapetus), five moons of Uranus (Miranda, Ariel, Umbriel, Titania, and Oberon), and one moon of Pluto (Charon). Some people refer to these moons as "satellites" of the planet they orbit.

Fig. 1.4 This strange looking object is of Phobos—one of the two moons of the planet Mars. The crater Stickney (*lower right*) dominates this side of Phobos (Credit: NASA).

Features of the Solar System

The Sun, planets, dwarf planets and their moons form a family of bodies called the **solar system**. We now know that the Sun is indeed at the centre of the solar system and that the major planets orbit the Sun in nearly circular orbits. Our understanding of the solar system has changed dramatically over the centuries as bigger and better telescopes were developed and more data of planetary motions was collected. Although people knew the planets of the solar system existed, little was known about the nature of these worlds until space probes containing scientific instruments were sent to explore these objects.

The Sun fits the definition of a star because it emits its own heat and light through the process of **thermonuclear fusion**.

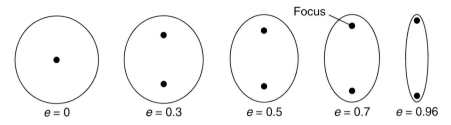

$e = 0$ $e = 0.3$ $e = 0.5$ $e = 0.7$ $e = 0.96$

Fig. 1.5 Shapes of planetary orbits (e = eccentricity).

Planets differ to stars in that they do not emit their own heat and light because they do not have enough mass for fusion to occur in their core. We see planets because they reflect sunlight. Stars and planets are spherical in shape and are held in that shape by gravity.

The planets orbit the Sun in much the same plane, and because of this they all appear to move across the sky through a narrow band of constellations called the zodiac. Observed from a position above the Sun's north pole, all the planets orbit the Sun in an anti-clockwise direction. The orbits of the planets in the solar system are often described as elliptical, but for most of the planets these ellipses are close to being circular. **Eccentricity** (e) is a measure of how far an orbit deviates from circularity. If e = 0 then the orbit if a perfect circle; while e becomes more elliptical as it approaches 1 (see Fig. 1.5).

The time taken by a planet to orbit the Sun is called its **period of revolution** and is also the length of its year. A planet's year depends on its distance from the Sun: the further a planet is from the Sun, the slower its speed and the longer its year.

As planets orbit the Sun, they also spin or rotate on an axis, which is an invisible line through their centre, from their north to south pole. A planet's **rotation period** is known as its day. Planets also have varying degrees of axial tilt. **Axial tilt** is the angle between a planet's axis of rotation and the vertical (Fig. 1.6).

Each planet has its own **gravitational field**, which tends to pull objects towards its centre. The strength of a gravitational field is measured in newtons per kilogram (N/kg) at the surface of a planet. It takes a lot of energy to overcome gravity and escape from the surface of a planet. The minimum speed that an object (such as a rocket) must attain in order to travel into space from the surface

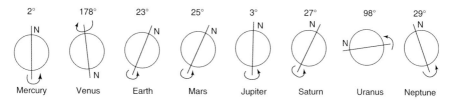

Fig. 1.6 Approximate axial tilt of each planet.

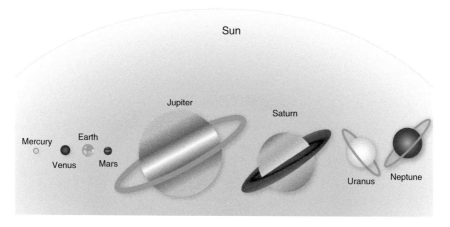

Fig. 1.7 The planets compared to the size of the Sun, with all bodies drawn to the same scale.

of a planet, moon or other body is called the escape velocity. If the rocket's velocity is too low, gravity will pull it back down.

Planets also have a **density**, which is a measure of the amount of mass in a given volume. Density is measured in kilograms per cubic metre (kg/m^3) but is often quoted in grams per cubic centimeter (g/m^3) as well. The density of water is 1.0 g/cm^3 while that of iron is 7.87 g/cm^3.

Planets often have a **magnetic field**. Such fields are produced by churning motions of metallic liquids in a planet's core that conduct electricity and have an electric charge. The magnetic fields act like a giant bar magnet and can be offset from the rotation axis of a planet. For example, the Earth's magnetic field is tilted about $11°$ to its axis of rotation.

The liquid conducting material in a planet's interior can be made to swirl about if the planet is rotating quickly enough. The faster a planet rotates, the more the material gets stirred up and the

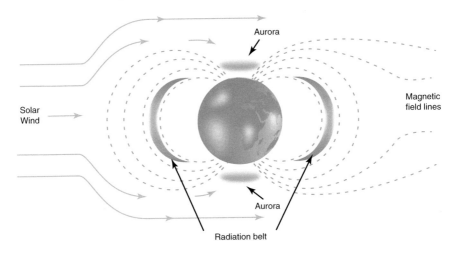

Fig. 1.8 The magnetic field around Earth. Charged particles in the solar wind can get trapped in the magnetic field near the poles and create 'aurora'. Some charged particles also get trapped in the Van Allen Radiation Belts.

stronger the generated magnetic field. If the liquid interior becomes solid or if the rotation slows down, the magnetic field will weaken.

Magnetic fields protect a planet from the charged particles streaming out from the Sun in the form of the solar wind. When solar wind particles run into a magnetic field, they are deflected and spiral around the magnetic field lines. Most solar wind particles are deflected past a planet, but a few leak into the magnetosphere to get trapped in radiation belts that surround the planet. Some particles gain enough energy to interact with atoms and molecules in the atmosphere of the planet to create a light show of **auroras** (see Fig. 1.8).

The four 'inner planets' (Mercury, Venus, Earth and Mars) are called **terrestrial planets**. They are smaller, denser and rockier than the 'outer planets' (Jupiter, Saturn, Uranus, and Neptune). The inner planets are also warmer and rotate more slowly than the outer planets. The **outer planets** are gaseous planets, containing mostly hydrogen and helium with some methane and ammonia. These rapidly rotating planets are cold and icy with deep atmospheres.

During the 1800s, astronomers discovered a large number of small, rocky bodies orbiting the Sun between Mars and Jupiter. Bodies such as Ceres, Pallas and Vesta, which had been thought of as small planets for almost half a century, became classified as **asteroids**.

Formation of the Solar System

The solar system is thought to have formed about 4.5 billion years ago from a vast cloud of very hot gas and dust called the **solar nebula**. This cloud of interstellar material began to condense under its own gravitational forces. As a result, density and pressure at the centre of the nebula began to increase, producing a dense core of matter called the protosun. Collisions between the particles in the core caused the temperature to rise deep inside the protosun.

The planets and other bodies in the solar system formed because the solar nebula was rotating. Without rotation, everything in the nebula would have collapsed into the protosun. The rotating material formed a flat disc with a warm centre and cool edges. This explains why nearly all the planets now rotate in much the same plane.

As the temperature inside the protosun increased, light gases like hydrogen and helium were forced outward while heavy elements remained closer to the core. The heavier material condensed to form the inner planets (which are mainly rock containing silicates and metals), while the lighter, gaseous material (methane, ammonia and water) condensed to form the outer planets. Thus a planet's composition depends on what material was available at different locations in the rotating disc and the temperature at each location (Fig. 1.9).

The formation of the solar system took millions of years. During this time the temperature and pressure of the protosun continued to increase. Finally the centre of the protosun became hot enough for nuclear fusion reactions to begin and the Sun was born.

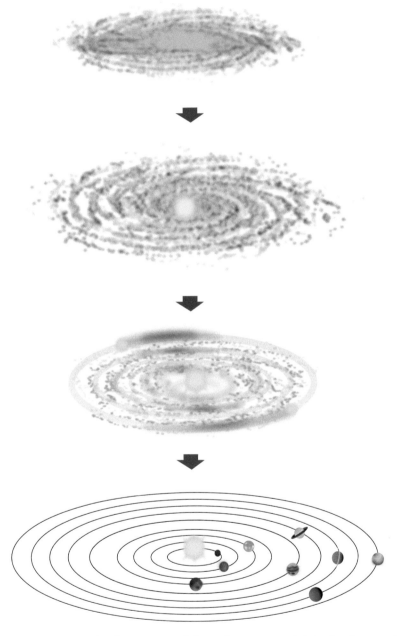

Fig. 1.9 Stages in the formation of the solar system: (a) A slowly rotating cloud of interstellar gas and dust begins to condense under its own gravity. (b) A central core begins to form a protosun. A flattened disc of gas and dust surrounds the protosun, and begins to rotate and flatten. (c) The planets begin to condense out of the flattened disc as it rotates. (d) The planets have cleared their orbit of debris.

Stars like our Sun can take 100 million years to form from a nebula. Radioactive data of the oldest material in our solar system suggests it is about 4.6 million years old.

Much of the debris leftover from the formation of the solar system is in orbit around the Sun in two regions—the Asteroid belt and the Kuiper belt. The Asteroid belt lies between Mars and Jupiter while the Kuiper belt is a region beyond Neptune.

The Asteroid Belt

Asteroids are small rocky boulders left over after the solar system formed. Most of them orbit the Sun in a region between the planets Mars and Jupiter. This region is called the Asteroid belt. The Asteroid belt is not uniformly thick, but is instead tapered. On its outer skirts the thickness is about 1 AU, but on it's inside edge the belt is only one-third as deep. Most asteroids have very elliptical orbits, some of which actually cross Earth's orbit. Small stony fragments ejected during asteroid collisions are called meteoroids. Some of these meteoroids enter the Earth's atmosphere and burn up releasing light—such bodies are called meteors. Large meteors that impact with the ground form craters like those seen on the surface of the Moon. More in Chap. 8.

The Kuiper Belt

The Kuiper belt is a bit like the Asteroid belt, except that it is much farther from the Sun and it contains thousands of very cold bodies made of ice and rock. Objects in this outer region take up to 200 years to orbit the Sun. The Kuiper belt is a remnant of the original solar nebula out of which our planetary system formed, but it is made up of material that could not coalesce into a planet due to the large volume of space and low density of matter so far from the Sun. Although the main Kuiper belt ends around 48 AU from the Sun, there is another large reservoir of objects called the Oort cloud, in the distant solar system (see Chap. 13).

The Oort Cloud

There are other smaller objects on the outer edge of solar system, in a region called the Oort cloud. This roughly spherical cloud also contains many objects left over from the formation of the solar system. This cloud extends over one-third of the way to the nearest star system, Alpha centauri, or about 100,000 AU. It likely contains around a trillion objects larger than 1 km across, with orbital periods of a few million years.

There are three competing theories for how the inner Oort cloud might have formed. One theory is that a rogue planet could have been tossed out of the giant planet region and could have perturbed objects out of the Kuiper belt to the inner Oort cloud on its way out. This planet could have been ejected or still be in the distant solar system today. The second theory is that a close stellar encounter could put objects into the inner Oort cloud region. A third theory suggests inner Oort cloud objects are captured extra-solar planets from other stars that were near our Sun in its birth cluster. More in Chap. 13.

Comets

Comets are icy bodies that originate from the outer regions of the solar system. Many comets have highly elongated orbits that occasionally bring them close to the Sun. When this happens the Sun's radiation vaporizes some of comet's icy material, and a long tail is seen extending from the comet's head. Each time they pass the Sun, comets lose about 1 % of their mass. Thus comets do not last forever. Comets eventually break apart, and their fragments often give rise to many of the meteor showers we see from Earth (see Chap. 13, Fig. 1.10).

Fig. 1.10 Comet ISON shows off its tail in this 3-min exposure taken on 19 Nov 2013, using a 14-inch telescope located at the Marshall Space Flight Center. The star images are trailed because the telescope is tracking on the comet. At the time of this image, Comet ISON was some 70 million km from the Sun—and 128 million km from Earth—moving at a speed of 217,000 km/h (Credit: NASA/MSFC/Aaron Kingery).

The Modern Nebula Theory

The Modern Nebula theory is supported by observations of dense dusty discs around very young stars in certain nebulas in the universe. It is believed that planets originate in these dense discs. The density of these discs has to be sufficient to allow the formation of the planets and yet be thin enough for the residual matter to be blown away by the star as its energy output increases. In 1992, the Hubble Space Telescope (HST) obtained the first images of these proto-planetary discs (called 'proplyds') in the Orion nebula. The Orion nebula star-birth region is 1500 light-years away, in the direction of the constellation Orion the Hunter. Some of the Orion proplyds are visible as silhouettes against a background of hot, bright interstellar gas, while others are seen to shine brightly.

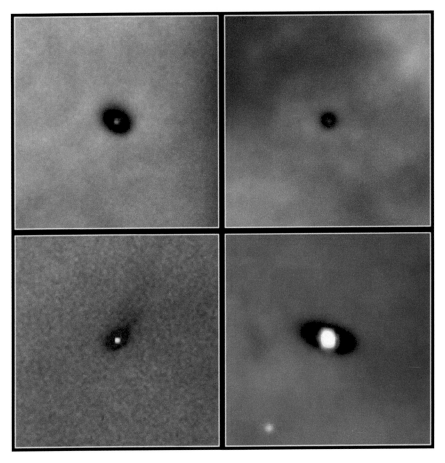

Fig. 1.11 Protoplanetary discs or proplyds in the Orion nebula as seen by the Hubble Space Telescope (Credit: NASA/ESA).

They are roughly on the same scale as our solar system and lend strong support to the nebular theory of their origin (see Fig. 1.11).

The Modern Laplacian Theory

Planetary mathematician Dr. Andrew Prentice of Monash University, Australia, has found widespread acclaim following the success of his many key predictions about the physical and chemical structure of the solar system. These predictions are linked to his Modern Laplacian theory of how the solar system was formed some 4.5 billion years ago. The controversial theory, first presented by Dr. Prentice in 1976, is based on a hypothesis advanced by French mathematician Pierre de Laplace in 1796.

Laplacian theory proposes that when the Sun first formed, it was a huge swirling cloud of gas and dust. When this cloud contracted inwards to form the present Sun, it cast off a concentric family of orbiting gas rings. The planets later condensed from these rings, starting with Neptune and finishing with Mercury. For the Sun to shed gas rings, Dr. Prentice introduced a new physical concept of 'supersonic turbulence'. It is this phenomenon that causes the primitive Sun to shed individual gas rings, one at the orbit of each planet.

Prentice has made a long list of controversial predictions about the nature of our solar system. To the surprise of many of his colleagues, recent NASA missions have confirmed that many of his hypotheses are remarkably accurate. Lets look at some of these predictions.

- In 1977, Prentice hypothesized that a rocky moon belt existed at four planetary radii from Jupiter's centre. Two years later, such a rocky ring was discovered, though closer to Jupiter than Prentice had predicted. He also predicted that Uranus had two more moons or moonlet streams than commonly thought. Nine years later, a new moon (Puck), was discovered to be orbiting Uranus, in addition to a family of nine moonlets.
- In 1980, Prentice predicted that Titan was not a native moon of Saturn but instead had been captured soon after Saturn had formed.
- In 1981, Prentice theorized that the mass of Saturn's moon Tethys was in fact 20–25 % larger than the generally predicted level. Three months later, it was confirmed to be 21 % larger than previously thought.
- In 1989, Prentice predicted that Neptune had four additional dark moons, at 5, 3.5, 2.5 and 1.8 radii in Neptune's equatorial plane. By the end of the year, four dark moons were discovered in Neptune's equatorial plane at 7, 3, 2.5 and 2.1 radii. He also predicted that dry ice would be the main carbon-bearing chemical on Triton. Three years later, infrared devices confirmed this.

- Prentice also correctly predicted Jupiter's outermost Galilean satellite, Callisto, was a cold, magnetically inert body of rock and ice, and that the smallest Galilean moon, Europa, would have a 150 km deep mantle of ice. The sulphur content of Jupiter's atmosphere was also found to have exactly matched Prentice's prediction.
- Data from NASA's space probe missions also proved Prentice's 25-year old theory that Jupiter's fifth largest moon, Amalthea, discovered in 1892, was actually a 'captured' asteroid and not a native satellite or moon of Jupiter.

The Nice Model

The Nice model of the solar system is a set of theories in which the orbits of the giant planets changed long after the planets formed. This relatively recent theory (2005) proposes that the planets Jupiter, Saturn, Uranus and Neptune originally had near circular orbits and were closer together than in the present. Objects (called planetesimals) in the Asteroid belt and Kuiper belt were also in slightly different positions—these objects slowly leaked out of their original positions and many were gravitationally scattered by the giant planets. The orbits of the giant planets altered quickly and dramatically—Jupiter was moved slightly inwards while Saturn, Uranus and Neptune were moved outwards. The resulting planetary rearrangement unleashed a flood of comets and asteroids throughout the solar system. Some of the planetesimals were thrown into the inner solar system, producing a sudden influx of impacts on the terrestrial planets including Earth's Moon (a time called the Late Heavy Bombardment). Eventually, the giant planets reached their current positions, and dynamical friction with the remaining planetesimal disc dampened their eccentricities and made the orbits of Uranus and Neptune circular again.

The Nice model is favored for its ability to explain the movement and position of objects in the Kuiper belt. As Neptune migrated outward, it came closer to the objects in the proto-Kuiper belt, capturing some of them and sending others into chaotic orbits. The objects in the Scattered disc and Oort cloud are believed to have been placed in their current positions by interactions with Neptune and Jupiter.

The Grand Tack Hypothesis

The Grand Tack hypothesis (published 2011) proposes that strong currents of flowing gas moved Jupiter inwards in the earliest 1–10 million years of the solar system, fundamentally altering the orbits of the asteroids and other planets. At one point it was positioned close to where the orbit of Mars is now. The planet's movements changed the nature of the asteroid belt and caused Mars to be smaller than it should have been. Like Jupiter, Saturn also got drawn towards the Sun. Gradually all the gas and dust in between Jupiter and Saturn got expelled, bringing their sun-bound death spiral to a halt and eventually reversing the direction of their motion. The two planets moved outwards over millions of years until they reached their current positions.

Juipter's movement helps explain why the asteroid belt is made up of both dry, rocky objects and icy objects. Astronomers think that the asteroid belt exists because Jupiter's gravity prevented the rocky material there from coming together to form a planet, instead remaining as a loose collection of objects. As Jupiter moved away from the Sun, the planet nudged the asteroid belt back inward and into its present position. Jupiter also deflected some of the icy objects in the outer regions of the solar system, into the inner parts.

In Conclusion

There have been many attempts to develop theories for the origin of the solar system. None of them can be described as totally satisfactory and it is possible that there will further developments which may better explain the known facts. However the majority of astronomers do believe that the Sun and the planets formed from the contraction of part of a gas/dust cloud under its own gravitational pull and that the small net rotation of the cloud was responsible for the formation of a disc around the central condensation. The central condensation eventually formed the Sun while small condensations in the disc formed the planets and their satellites. The energy from the young Sun blew away the remaining gas and dust leaving the solar system as we see it today.

Further Information

https://solarsystem.nasa.gov (click on What is a planet? and Our solar system)

www2.ess.ucla.edu/~jewitt/kb/nice.html (The Nice model)

www.solarsystem.nasa.gov/scitach/display.cfm?ST_ID2429 (Grand Tack model)

www.boulder.swri.edu/~kwalsh/GrandTack.html (Grand Tack model)

2. Space Probes and Telescopes

Highlights

- The Hubble Space Telescope is one of the most significant instruments ever used to explore the solar system and universe.
- The Chandra space probe discovered X-rays coming from Jupiter's poles.
- The Spitzer space probe discovered the dwarf planet Makemake in the outer solar system.
- In 2005 the Deep Impact probe intercepted comet Tempel 1 and fired a 370 kg copper slug into the surface of the comet at 10 km/s.
- The New Horizons probe visited the dwarf planet Pluto in July 2015.
- In 2014, the Rosetta spacecraft encountered the comet Churyumov–Gerasimenko and placed a lander on it.
- The Dawn space probe is the first spacecraft ever to orbit two worlds beyond Earth (visiting the asteroid Vesta in 2011 and the dwarf planet Ceres in 2015).

Humans have long been interested in observing the night sky and in particular the planets and Moon. Early humans made observations with their unaided eyes. The invention of the telescope in 1610 opened up a new method to observe these objects. Telescope technology has improved dramatically since 1610 and we now have highly advanced optical telescopes as well as radio telescopes to find out more about the objects in the solar system.

The race to explore space advanced even more in the 1960s when the first artificial satellites where placed in Earth orbit. Since then we have seen advances in unmanned space probes and manned spacecraft through the USA's Mercury, Gemini and Apollo programs and the Russian Vostok, Voskhod, and Soyuz projects. It was largely as a result of these programs that the first

J. Wilkinson, *The Solar System in Close-Up*, Astronomers' Universe,
DOI 10.1007/978-3-319-27629-8_2,
© Springer International Publishing Switzerland 2016

Fig. 2.1 The International Space Station is a joint venture between several countries. The first part of it was put in orbit in 1998 and several modules have since been added. It now measures 108 m across and 88 m long, and has almost half a hectare of solar panels to power it. It orbits Earth at an altitude of 400 km (Credit: NASA/ISS).

human was able to walk on the Moon in 1969. This was perhaps the most significant step in the exploration of the Solar System, since the Moon was the first object outside Earth that humans ventured onto.

Since landing on the Moon, humans have turned their attention to improving methods of space travel and to studying the long-term impact of space travel on humans. As a result of this new focus, manned space stations like Salyut, Mir, Skylab and ultimately the International Space Station were built. Future manned missions into space will require new technologies and the current space stations are providing a pathway for this to occur.

Space Telescopes

In the past decade most observation of objects in the solar system has occurred through the use of unmanned space probes and space telescopes (telescopes that orbit the Earth as a satellite).

Table 2.1 Significant space telescopes (still operating in 2015)

Name	Operator	Main wavelength	Launch date	Orbits...
Hubble ST	NASA/ESA	UV/VIS	24 Apr 1990	Earth
Chandra	NASA	X-ray	23 Jul 1999	Earth
XMM-Newton	ESA	X-ray	10 Dec 1999	Earth
Odin	Swedish SC	Microwave	20 Feb 2001	Earth
Intern Astro. Lab	ESA	Gamma γ	17 Oct 2002	Earth
Spitzer	NASA	IR	25 Aug 2003	Sun
Swift Burst Expl.	NASA	UV, VIS, X, γ	20 Nov 2004	Earth
AGILE	ISA	X-ray	23 Apr 2007	Earth
Fermi	NASA	Gamma	11 Jun 2008	Earth
Kepler	NASA	UV, VIS	6 Mar 2009	Earth

Significant space telescopes include the Hubble Space Telescope (HST), the Chandra Space Telescope, the XMM-Newton Telescope and the Spitzer Space Telescope. The scopes are placed in orbit above Earth's surface to avoid the dusty and light affected atmosphere. Each of these telescopes observes objects in the solar system or universe in a particular wavelength of light. Some of the wavelengths are actually absorbed by the atmosphere and so we can't observe in these wavelengths from the surface (see Table 2.1).

The Hubble Space Telescope

The Hubble Space Telescope (HST) is one of the most significant instruments ever used to explore the solar system and universe. It was launched from the cargo bay of the shuttle Discovery on 25 April 1990 as a joint venture between NASA and the European Space Agency (ESA). The telescope cost $2.5 billion, weighs nearly 12 tonne and orbits 600 km above the Earth at a speed of 28,000 km/h. It consists of a 2.4-m-diameter mirror mounted in large tube, three cameras and two spectrographs and a number of guidance sensors. Hubble can detect objects about a billion times fainter than the human eye. Scientists are using the telescope to learn about the nature of stars, planets and black holes, the evolution of the universe, and distant objects never seen before (see Fig. 2.2).

During its lifetime, the HST has been studying the universe at wavelengths from the infrared through to the ultraviolet. Hubble has been used to view over 30,000 objects throughout the universe

Fig. 2.2 The Hubble Space Telescope (Credit: NASA/HST).

and recorded over 44 terabytes of data. Astronomers using the HST data have published over 8700 scientific papers. Its high-resolution images of Mars, Jupiter, Saturn and Neptune are providing surprising detail about these planets. The world was amazed in July 1994 when Hubble produced images showing the impact of the comet Shoemaker-Levy 9 on Jupiter. Hubble has been used to study the Great Nebula in Orion and it has detected dusty discs around protostars that are thought to be new solar systems forming. Discs of matter have also been seen swirling around super massive black holes at the centre of galaxies and quasars, as well as structure in the spiral arms of nearby galaxies.

The Chandra Space Telescope

The Chandra space telescope was launched by the shuttle Columbia in July 1999, and is designed to observe X-rays from high-energy regions of the universe, such as the remains of exploded stars. X-rays provide scientists with a different perspective when exploring space. Chandra discovered that Jupiter's X-rays are coming from its poles and not the auroral rings on the planet.

Chandra has 8 times greater resolution and is able to detect sources more than 20 times fainter than any other previous X-ray telescope. Chandra's highly elliptical orbit allows it to observe continuously for up to 55 h of its 65 h orbital period. At its furthest orbital point from Earth, Chandra is one of the most distant Earth-orbiting satellites. This orbit takes it beyond the geostationary satellites and beyond the outer Van Allen belt.

Although Chandra was initially given an expected lifetime of 5 years, on 4 September 2001 NASA extended its lifetime to 10 years based on the observatory's outstanding results. Physically Chandra could last much longer. A study performed at the Chandra X-ray Centre indicated that the observatory could last at least 15 years.

The XXM-Newton Space Telescope

The XMM-Newton telescope was launched by rocket in December 1999, and is designed to investigate the origins of the universe by probing cosmic matter from black holes. It has detected X-ray emission from solar system objects and is also used to study star forming regions in the universe. XMM-Newton orbits Earth in an elliptical orbit between 7000 and 114,000 km above Earth. The satellite weighs 3800 kg and is 10 m long and is 16 m in span with its solar arrays deployed. It holds three X-ray telescopes, developed by Media Lario of Italy, each of which contains 58 Wolter-type concentric mirrors. The combined collecting area is 4300 cm^2. The three European Photon Imaging Cameras are sensitive over the energy range 0.2–12 keV. Other instruments onboard are two reflection-grating spectrometers that are sensitive to below 2 keV, and a 30 cm diameter Ritchey-Chrctien optical/UV telescope.

The Spitzer Space Telescope

The Spitzer Space Telescope (also known as the Space Infrared Telescope Facility or SIRTF) was launched in August 2003 and observes mainly in the infrared region. Infrared rays are blocked by the Earth's atmosphere, so observation is only possible from space at this wavelength. This telescope orbits Earth at a height of 568 km and is used to study asteroids in the solar system, gas giant planets, dusty stars and distant galaxies. In 2005 Spitzer was used to discover the dwarf planet Makemake in the outer solar system.

Instead of orbiting Earth like other space telescopes, Spitzer tags along behind Earth as it orbits the Sun. This keeps it clear of Earth's heat and makes for better pictures. Spitzer will be able to see every part of the sky at least every 6 months during its life.

The Kepler Space Telescope

Kepler is a space observatory launched by NASA to discover Earth-like planets orbiting other stars. The spacecraft has a mass of 1039 kg and contains a 1.4-m primary mirror feeding an aperture of 0.95-m—at the time of its launch this was the largest mirror on any telescope outside of Earth orbit. The focal plane of the spacecraft's camera is made up of 42 CCDs at 2200×1024 pixels, which made it at the time the largest camera yet launched into space, possessing a total resolution of 95 megapixels. Heat pipes connected to an external radiator cool the array.

The spacecraft is named after the Renaissance astronomer Johannes Kepler and was launched on 7 March 2009. Kepler's sole instrument is a photometer that continually monitors the brightness of over 145,000 stars in our region of the Milky Way galaxy. This data is transmitted to Earth, then analysed to detect periodic dimming caused by exoplanets that cross in front of their host star. As of February 2014, Kepler and its follow-up observations had found over 900 confirmed exoplanets in more than 76 stellar systems, along with a further 2903 unconfirmed planet candidates. In November 2013, astronomers reported that there could be as many as 40 billion Earth-sized planets orbiting in

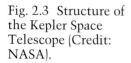

Fig. 2.3 Structure of the Kepler Space Telescope (Credit: NASA).

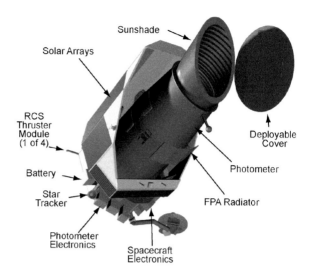

the habitable zones of Sun-like stars and red dwarf stars within the Milky Way Galaxy.

Kepler orbits the Sun, which avoids Earth occultations, stray light, and gravitational perturbations and torques inherent in an Earth orbit. Kepler's photometer points to a field in the northern constellations of Cygnus, Lyra and Draco, which is out of the ecliptic plane, so that sunlight never enters the photometer as the spacecraft orbits the Sun.

In April 2012, an independent panel of senior NASA scientists recommended that the Kepler mission be continued through 2016. According to the senior review, Kepler observations needed to continue until at least 2015 to achieve all the stated scientific goals (see Fig. 2.3).

Future Space Telescopes

There are a number of space telescopes panned to be launched in the near future. These include LISA Pathfinder (2015), James Webb ST (2018), Hard X-ray Modulation Telescope (2014–2016) and Dark Matter Particle Explorer (2015–2016).

The **James Webb Space Telescope** (JWST), previously known as Next Generation Space Telescope, is a planned space telescope designed for observations mainly in the infrared region of the spectrum. It is to be the scientific successor to the Hubble Space

Telescope and the Spitzer Space Telescope. The main technical features are a large and very cold 6.5-m diameter mirror and four specialised instruments at an observing position far from Earth, orbiting the Earth–Sun L2 point. The combination of these features will give the telescope unprecedented resolution and sensitivity from long-wavelength visible to the mid-infrared, enabling its two main scientific goals—studying the birth and evolution of galaxies, and the formation of stars and planets. JWST's instruments will not measure visible or ultraviolet light like the Hubble Telescope, but will have a much greater capacity to collect infrared light.

The JWST telescope is a project of the National Aeronautics and Space Administration (NASA, the United States space agency), with international collaboration from the European Space Agency (ESA) and the Canadian Space Agency (CSA), including contributions from 15 nations. The launch date is 2018 and the mission duration is 5 years minimum with a possibility of 10 years (see Fig. 2.4).

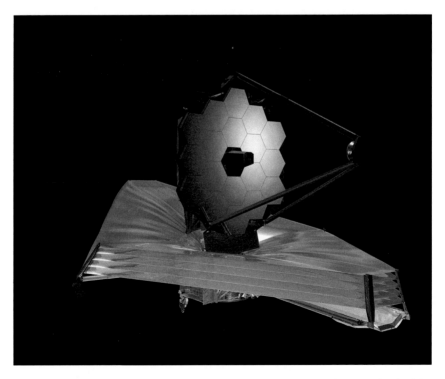

Fig. 2.4 Artists picture of what the James Webb Space Telescope looks like in space (Credit: NASA).

Using Space Probes to Explore the Solar System

Humans have always had the vision to 1 day live on other planets. This vision existed even before the first person was put into orbit. Since the early space missions of putting humans into orbit around Earth, many advances have been made in space technology. It is now possible to send unmanned space probes deep into the Solar system to explore other planets. Humans have only travelled as far as the Moon and back, but robotic space probes have been placed on the surface of planets like Venus and Mars, as well as on the surface of an asteroid. Probes have also been used to orbit distant planets like Jupiter and Saturn.

Planetary probes travel in large orbits around the Sun. They often pass target planets (flybys), go into orbit around planets or land on planets. Instruments onboard space probes collect information about the planet and return it via radio signals back to Earth.

The first planet to which humans sent a spacecraft was Venus, the closest planet to Earth. Venus is similar in size and mass to Earth and has always been of interest to humans. Mars was the next planet to which a probe was sent, followed by Mercury. Some of these early space probes were successful while others failed—all however provided information and experiences that lead to future success.

Current Probes in the Solar System

There are a number of significant space probes currently in operation exploring the solar system. Details of these probes and their missions will be discussed in the relevant chapters in this book. However, several recent probes are significant enough to deserve special mention here.

The Messenger Probe

Messenger (MErcury Surface, Space Environment, Geochemistry, and Ranging) is a robotic space probe launched by NSSA in August 2004 to study the planet Mercury. Messenger became the second mission after Mariner in 1975 to visit Mercury. Messenger made two flybys of the planet in 2008 and finally entered orbit around Mercury in March 2011.

The primary mission was completed in March 2012 after the entire surface of the planet was mapped and 100,000 images were recorded. An extension of the program was made in July 2013. The probe made a final orbital correction in January 2015 leading to a termination of the mission with an impact into Mercury's northern hemisphere in April 2015.

Instruments on Messenger have collect valuable data on Mercury and its environment.

Its surface looks much like that of Earth's Moon.

Messenger has a dual-mode, liquid chemical propulsion system that is integrated into the spacecraft's structure to make economical use of mass. The structure is primarily composed of a graphite epoxy material. This composite structure provides the strength necessary to survive launch while offering lower mass. Two large solar panels, supplemented with a nickel-hydrogen battery, provide Messenger's power (see Fig. 2.5).

The New Horizons Probe

The New Horizon probe was launched by NASA on 19 January 2006 on a mission to study the dwarf planet Pluto and the Kuiper belt (see Figs. 2.6 and 2.7). In July 2015, the craft was the first to fly by Pluto and its moons. Radio signals took over 4 h to travel from the craft back to Earth. The craft set the record for the highest velocity of a human-made object from Earth at 58,536 km/h. It flew by the orbit of Mars on 7 April 2006, Jupiter on 28 February 2007, the orbit of Saturn on 8 June 2008 and the orbit of Uranus on 18 March 2011. New Horizons flew within 10,000 km of Pluto with a velocity of 49,600 km/h. It also came as close as 27,000 km

Fig. 2.5 The Messenger probe is exploring the surface of Mercury (Credit: NASA/JHU/CIW).

to Charon (Pluto's largest moon). The spacecraft's instrumentation includes a high resolution telescope, another scope with broadband spectroscopic capacity going into the near infrared and far ultraviolet, a particle and electron detector, a dust counter and radio science experiments using the communication channels.

New Horizons was originally planned as a voyage to what was the only unexplored planet in the Solar System—Pluto. When the spacecraft was launched, Pluto was still classified as a planet, later to be reclassified as a dwarf planet by the International Astronomical Union (IAU). Some members of the New Horizons team disagreed with the IAU definition and still describe Pluto as the ninth planet. Pluto's satellites Nix and Hydra also have a connection with the spacecraft: the first letters of their names ("N" and "H") are the initials of "New Horizons". The moons' discoverers chose these names for this reason, in addition to Nix and Hydra's relationship to the mythological Pluto.

Fig. 2.6 The New Horizons space probe was launched into orbit via an Atlas V rocket from Cape Canaveral Air Force Station in Florida, USA on 19 January 2006 (Credit: NASA).

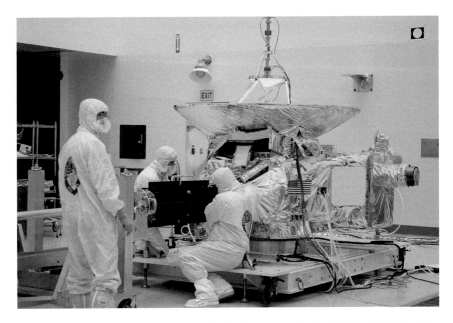

Fig. 2.7 New Horizon's space probe in its assembly hall (Credit: NASA).

Pluto was discovered by Clyde Tombaugh in 1930. About an ounce of Tombaugh's ashes are aboard the New Horizon's spacecraft, to commemorate his discovery. A Florida-state quarter coin, whose design commemorates human exploration, is also included. One of the science packages (a dust counter) is named after Venetia Burney, who, as a child, suggested the name "Pluto" after the planet's discovery.

After passing by Pluto, New Horizons will continue farther into the Kuiper belt. Mission planners are now searching for one or more additional Kuiper belt objects (KBOs) of the order of 50–100 km in diameter for flybys similar to the spacecraft's Plutonian encounter.

The Stereo Probe

The STEREO (Solar TErrestrial RElations Observatory) probe is a solar mission launched by NASA on 26 October 2006. It consists of two nearly identical spacecraft, one orbiting ahead of Earth (A) and the other behind Earth (B). Observations are made simultaneously of the Sun and then combined to provide a 3-D stereo image of the Sun. Spacecraft A takes 347 days to orbit the Sun while spacecraft B takes 387 days. Because the A spacecraft is moving faster than B, they are separating from each other and A is orbiting closer to the Sun than B. The images are adjusted to account for this difference.

STEREO is used to image the inner and outer corona and the space between Sun and Earth, detect electrons and other energetic particles in the solar wind, study the plasma characteristics of protons, alpha particles and heavy ions, and monitor radio wave disturbances between the Sun and Earth.

From February 2011, the two Stereo spacecraft will be 180° apart from each other, allowing the entire Sun to be seen for the first time. Such observations will continue for several years. By combining images from the STEREO A and B spacecraft, with images from NASA's Solar Dynamic Observatory (SDO) satellite, a complete map of the Sun can be formed. Previous to the STEREO mission, astronomers could only see the side of the Sun facing Earth, and had little knowledge of what happened to solar features after they rotated out of view. In 2015 contact with the two

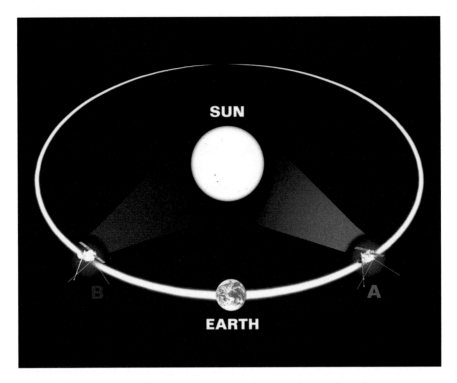

Fig. 2.8 Positions of the two Stereo probes as they orbit the Sun. From February 2011 the two craft will be 180° apart from each other, allowing the entire Sun to be seen in stereo (Credit: NASA).

spacecraft was temporarily lost for a few months as they both passed behind the Sun. Contact was re-established with Stereo A but not with Stereo B (see Fig. 2.8).

The Rosetta Probe

Rosetta is a robotic spacecraft built by the European Space Agency (ESA) to study the comet 67P/Churyumov–Gerasimenko. The probe was launched on March 2004 on an Ariane 5 rocket and reached the comet in May 2014. The spacecraft consists of two main elements: the Rosetta space probe orbiter, which features 12 instruments, and the Philae robotic lander, with an additional nine instruments. Rosetta will orbit the comet for 17 months. The spacecraft has already performed two successful asteroid flyby missions on its way to the comet. In 2007 Rosetta also performed a flyby of Mars and returned useful images. The craft completed its flyby of asteroid 2867 in September 2008 and of 21 Lutetia in July 2010. On 20 January 2014, Rosetta was taken out of a 31-month hibernation mode and is continuing to its target.

In May 2014, the Rosetta spacecraft entered a slow orbit around the comet and gradually slowed down before releasing the lander in November 2014. The lander approached the comet at speed around 3.6 km/h and on contact with the surface, two harpoons were to be fired into the comet to prevent it from bouncing off. However the harpoons didn't fire and the lander bounced back up a kilometer into space, soaring for nearly 2 h before returning to the ground. After another small bounce, the lander settled somewhere in the shadow of a cliff, about a kilometre from where it was supposed to be. The lander was able to collect data with its suite of instruments, sniffing, hammering, drilling, and even listening to the comet (see Chap. 13).

The Dawn Probe

NASA launched the Dawn space probe on 27 September 2007 on a mission to study the two largest objects in the asteroid belt—Vesta and Ceres. The spacecraft uses ion propulsion to transverse space far more efficiently than if it used chemical propulsion. In an ion propulsion engine, an electrical charge is applied to xenon gas, and charged metal grids accelerate the xenon particles out of the thruster. These particles push back on the thrusters as they exit; creating a reaction force that propels the spacecraft forward. Dawn has now completed over 5 years of accumulated thrust time, far more than any other spacecraft.

Dawn was the first spacecraft to enter orbit around Vesta on 16 July 2011. The probe gave scientists a much closer view of this object. Dawn was originally scheduled to depart Vesta and begin its journey to Ceres on 26 August 2012, however, a problem with one of the spacecraft's reaction wheels forced Dawn to delay its departure from Vesta's gravity until 5 September 2012. Dawn successfully departed Vesta on this date and arrived at Ceres in April 2015, 3–4 months prior to the arrival of New Horizons at

Pluto. Dawn is the first mission to study a dwarf planet at close range. It's mission calls for it to enter orbit around Ceres at an initial altitude of 13,500 km for a first full characterisation. Dawn will then spiral down to a survey orbit at an altitude of 4430 km.

Solar Dynamics Observatory

The Solar Dynamics Observatory (SDO) is the most advanced spacecraft ever designed to study the Sun and its dynamic behavior. SDO is providing better quality, more comprehensive science data faster than any NASA spacecraft currently studying the Sun. The probe is aimed at providing data on the processes inside the Sun, the Sun's surface, and its corona that result in solar variability. SDO will help scientists to better understand the Sun's influence on Earth and near-Earth space through the use of many wavelengths simultaneously. SDO is investigating how the Sun's magnetic field is generated and structured.

SDO was launched from Cape Canaveral Air Force Station in the USA on 11 February 2010. After launch, SDO was placed into an orbit around Earth at an altitude of about 2500 km. It then underwent a series of orbit-raising maneuvers that placed it in a circular, geosynchronous orbit at altitude 36,000 km. It had a 5-year science mission but carries enough fuel to operate for an additional 5 years. At launch its mass was 3100 kg with a payload of 290 and 1450 kg of fuel. The solar panels cover an area of 6.6 m^2 producing 1450 W of power. The overall length of the spacecraft along the Sun-pointing axis is 4.5 m, and each side is 2.22 m (see Fig. 2.9).

The data from SDO is providing a lot of new information and spectacular images of the Sun. Scientists are gaining a better understanding of how even small events on the Sun can significantly effect the operation of technological infrastructure on Earth (such as GPS systems, cable TV, radio and satellite communications).

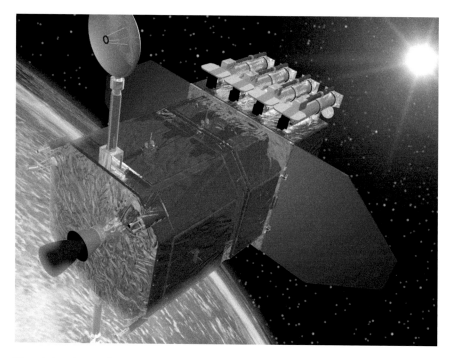

Fig. 2.9 The Solar Dynamics Observatory (SDO) is the most advanced spacecraft ever designed to study the Sun (Credit: NASA/SDO).

The Juno Probe

Juno is a spacecraft launched by NASA on 5 August 2011 on a mission to explore the planet Jupiter. The craft is expected to arrive at Jupiter in July 2016. It will be placed in a polar orbit to study the mass distribution, atmosphere and composition of Jupiter, as well as its gravitational and magnetic field. Juno will take 5 years to reach Jupiter. The probe's trajectory used a gravity assist speed boost from Earth, accomplished by two Earth flybys (one in October 2013 and the second in August 2011). In August 2016 the spacecraft will perform an orbit insertion burn to slow the spacecraft enough to allow capture into an 11 day polar orbit. Once Juno enters orbit around Jupiter its infrared and microwave instruments will begin to measure the thermal radiation and convection currents within Jupiter's atmosphere. Juno's polar orbit is highly elliptical and takes it close to within 4300 km of the poles. This type of orbit helps the craft avoid any long-term contact with Jupiter's radiation belts, which can cause damage to spacecraft electronics and solar panels. The craft will complete at least 33 orbits, each taking from 11 to 14 days, before being crashed into Jupiter itself.

Mars Science Laboratory

NASA launched Mars Science Laboratory (MSL) on 26 November 2011 on a mission to Mars. The probe successfully landed a rover called Curiosity in Gale crater on 6 August 2012. The objectives of the Curiosity rover include investigating the possibility of life on mars, studying the climate and geology, and collecting data for any future manned missions to Mars.

Curiosity is about twice as long and five times as heavy as the Spirit and Opportunity rovers (already on the surface of Mars), and carries over ten times the mass of scientific instruments.

The MSL spacecraft that transported Curiosity to Mars successfully carried out a more accurate landing than previous rovers, within a landing ellipse of 7 by 20 km inside Gale crater. It is designed to explore the surface for at least 2 years covering an area of 5 km by 20 km. Curiosity has been able to drill into the surface of Mars and examine sediments formed by ancient river beds.

Probing Comets

A **comet** is an icy small body that originates in the Kuiper belt, Scattered Disc or Oort Cloud. Sometimes the orbit of a comet brings it into the inner solar system and near the Sun. When passing the Sun, comets heat up and begin to vaporise, emitting a visible tail of gaseous material.

Comets can tell astronomers much about the history of the Solar system, but only a few space probes have been used to explore the nature of comets.

In 2001, the Deep Space 1 spacecraft obtained high-resolution images of the surface of Comet Borrelly. The surface of this comet was found to be hot and dry, with a temperature of 26–71 °C and extremely dark. This suggested that the ice had been removed by solar heating and maturation, or was hidden by soot-like material.

NASA launched the Deep Impact space probe on 12 January 2005. The probe intercepted comet Tempel 1 in July 2005 and passed within 500 km of its nucleus. The probe fired a 370 kg copper slug into the surface of the comet at 10 km/s. Earth-based telescopes and space observatories, including Hubble, Chandra, Spitzer, and XMM-Newton, photographed the entire event. Cameras and spectroscopes on board the ESA's Rosetta spacecraft also observed the impact, which was about 80 million km from the comet at the time of impact. Rosetta determined the composition of the gas and dust cloud that was kicked up by the impact. However, the initial photographs taken of the impact site were unsatisfactory.

On 15 February 2011, NASA scientists identified the crater formed by the Deep Impact slug in images from the Stardust probe. The crater was estimated to be 150 m in diameter with a bright mound in the centre likely created when material from the impact fell back into the crater. The mission found that most of the comet's water ice is below the surface and that reservoirs feed the jets of vaporised water that form the body of the comet.

Since its encounter with the comet Tempel 1, Deep Impact has passed by other comets, including, Boethin (2008), Hartley 2 (2010), Garradd (2012) and comet ISON (2013).

During September 2013, Deep Impact mission controllers found the computers on the spacecraft were continuously rebooting themselves and so were unable to issue any commands to the vehicles thrusters. As a result of this problem, communication to the spacecraft became more difficult as the orientation of the vehicle's antennas was unknown. Additionally, the solar panels on the vehicle were positioned incorrectly for generating power. On 20 September 2013, NASA abandoned further attempts to contact the craft.

As previously mentioned the Rosetta probe encountered the comet Churyumov–Gerasimenko in 2014 and placed a small lander on its surface (more in Chap. 13).

Probes Leaving the Solar System

There are four probes that have passed the orbits of the major planets and are deemed to be leaving the solar system. These are, Pioneer 10 and Pioneer 11, and Voyager 1 and Voyager 2.

The United States began its exploration of more distant parts of the Solar System with its Pioneer space probes launched in 1972 and 1973. These craft were designed to survive the passage through the Asteroid Belt and Jupiter's magnetosphere. The Asteroid Belt was relatively simple to pass through since there were many gaps, but the space probes were nearly fried by ions trapped in Jupiter's magnetic field.

Launched in 1972, Pioneer 10 was the first spacecraft to fly by Jupiter in December 1973, passing within 130,000 km of the cloud-covered surface. Twenty-three low-resolution images were returned to Earth showing Jupiter's turbulent atmosphere and the Great Red Spot.

Pioneer 10's greatest achievement was the data collected on Jupiter's moons, its strong magnetic field, and interactions with the solar wind.

About a year later, Pioneer 11 flew to within 48,000 km of Jupiter's surface and sent back 17 images of the planet to ground crews on Earth. The space probe used the strong gravitational field of Jupiter to swing it on a path towards Saturn—a journey that was to take 5 years. In September 1979, Pioneer 11 passed within 30,000 km of Saturn's surface and it returned 440 images and data about Saturn's moons and its ring system.

Today Pioneer 10 and 11 are no longer sending back data, but both are still travelling at about 12 km/s and heading in opposite directions away from the solar system into deep space. Each craft carries a plaque, with a graphic message, to inform anyone out there about the Solar System, the Earth, and the human race. No further contact attempts are planned.

Voyager 1 and Voyager 2 were launched by the USA in 1977 on a mission towards Jupiter, the largest planet in the solar system. These probes flew by Jupiter in March and July of 1979 before proceeding to Saturn. Each Voyager was equipped with high-resolution cameras, three programmable computers and

Fig. 2.10 In order to track the Voyager spacecraft, NASA uses a system of deep space communication stations scattered across Earth's surface. One of these is located at Tidbinbilla near Canberra, Australia (Photo: J. Wilkinson).

instruments to conduct a range of scientific experiments. Voyager 2 was actually launched 16 days before Voyager 1 but Voyager 1 took a faster, more direct route to reach Jupiter first (Fig. 2.10).

On their journey through the solar system the Voyager probes discovered:

- 22 new satellites (3 at Jupiter, 3 at Saturn, 10 at Uranus and 6 at Neptune)
- Active volcanoes on Io
- An atmosphere and geysers on Triton
- Rings around Jupiter and spokes in Saturn's rings
- Auroral zones on Jupiter, Saturn and Neptune
- Large scale storms on Neptune
- Magnetospheres around Uranus and Neptune.

These two space probes provided many spectacular close-up views of the four outer planets known as the gas giants. Both space probes are still moving away from Earth at about 16 km/s and still in operation. They are returning data about cosmic rays in outer space and ultraviolet sources among the stars.

Both spacecraft also have adequate electrical power and attitude control propellant to continue operating until around 2025, after which there may not be available electrical power (from radioisotope thermoelectric generators) to support science instrument operation. At that time, science data return and spacecraft operations will cease. Interestingly, Voyager 1 has passed the Pioneer 10 space probe and is now the most distant human-made object in space.

Further Information

www.space.com/19081-solar-system-space-probes-missions.html
www.nineplanets.org/
www.hubblesite.org/
www.nasa.gov/mission_pages/hubble/main/
http://nssdc.gsfc.nasa.gov/ (check out solar system exploration)

3. The Dominant Sun

Highlights

- The STEREO and the Solar Dynamics Observatory are the most technologically advanced space probes used to study the Sun.
- Data obtained from the SOHO probe has shown that the strength of the magnetic fields around sunspots is thousands of times stronger than the Earth's magnetic field.
- The Ulysses space probe found that the solar wind blows faster around the Sun's poles (750 km/s) than in equatorial regions (350 km/s).
- The Japanese Hinode probe was the first to be able to measure small changes in the Sun's magnetic field.
- Solar probes have provided answers to many questions regarding the effect of the Sun on Earth's climate.

The Sun is the dominant object in the solar system because it is by far the largest object. It is positioned at the centre of the solar system and its gravitational pull holds all the planets in orbit. The Sun is an average sized star about 4.5 billion years old. Unlike planets, stars produce their own light and heat by burning fuels like hydrogen and helium in a process known as nuclear fusion. Stars have a limited life and the Sun is no exception—it is about half way through its life cycle of about 10 billion years.

The Sun is one of over 100 billion stars that make up a galaxy called the Milky Way. The Milky Way galaxy is spiral in shape, and the Sun is positioned about halfway out from the centre. The Milky Way is about 100,000 light years in diameter and 15,000 light years thick. You can see parts of the Milky Way as a band of

J. Wilkinson, *The Solar System in Close-Up*, Astronomers' Universe,
DOI 10.1007/978-3-319-27629-8_3,
© Springer International Publishing Switzerland 2016

cloud that stretches across the night sky. Within the Milky Way, the Sun is moving at 210 km/s, and takes 225 million years to complete one revolution of the galaxy's central mass of stars (see Fig. 3.1).

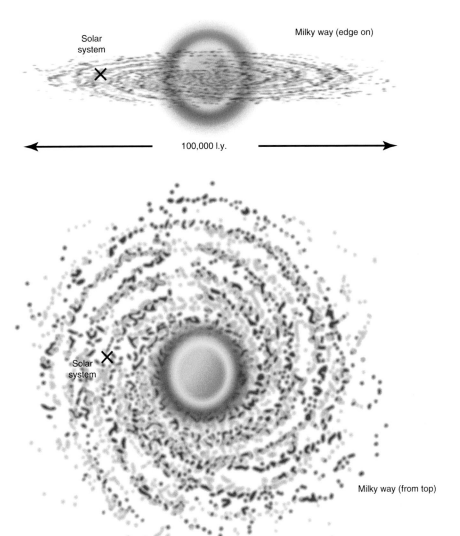

Fig. 3.1 Position of the Sun in the Milky Way galaxy.

Fig. 3.2 The STEREO (Ahead) spacecraft caught this spectacular eruptive prominence on the Sun (12–13th April 2010). The length of the prominence appears to stretch almost halfway across the Sun, about 800,000 km. Prominences are cooler clouds of plasma that hover above the Sun's surface, tethered by magnetic forces. They are notoriously unstable and commonly erupt as this one did in a dramatic fashion (Credit: NASA/STEREO).

Probing the Sun

Scientists have gained much of their knowledge about the Sun from observation made on Earth over many years. However, much of our current knowledge has come from space probes that have been sent on missions to investigate the Sun. These probes have provided accurate information about the Sun's temperature, atmosphere, composition, magnetic fields, flares, prominences, sunspots and internal dynamics.

The USA launched a number of unmanned solar probes between 1959 and 1968 as part of its Pioneer program. Many of these early probes have now completed their missions but still remain in orbit around the Sun. Missions such as Pioneer

Table 3.1 Recent probes used to observe the Sun

Spacecraft	Country of origin	Launch date	Mission focus
SOHO	USA/Europe	Dec 1995	Solar environment
TRACE	USA	April 1998	Magnetic field, corona
Genesis	USA	Aug 2001	Solar wind
Coronas F	Russia	July 2001	Flares and interior
RHESSI	USA	February 2002	X-ray imaging
Hinode	Japan, USA, UK	September 2006	Magnetic field
Stereo A/B	USA	October 2006	CMEs
SDO	USA	February 2010	Effect of Sun on Earth
SOLO	Europe	Due 2017	Study Sun up close
Solar probe plus	USA	Due 2018	Corona and solar wind

10 and 11 showed that gravity assists were possible and that spacecraft could survive high-radiation areas in space.

America's first space station, Skylab (launched in 1973), was used to study the Sun from Earth orbit. The space station included the Apollo Telescope Mount (ATM), which astronauts used to take more than 150,000 images of the Sun. Skylab was abandoned in February 1974 and re-entered the Earth's atmosphere in 1979. Table 3.1 lists the more recent probes that have been used to observe and explore the Sun.

Significant solar probes launched in the 1990s include Yohkoh (launched in 1991), Ulysses (October 1990), SOHO (December 1995), and TRACE (launched 1998). The most recently launched probes, STEREO and the Solar Dynamics Observatory (SDO), are also the most technologically advanced.

SDO was launched in 2010 and is designed to study the Sun and its dynamic behavior. This probe is providing better quality, more comprehensive science data faster than any NASA spacecraft currently studying the Sun. The probe is aimed at providing data on the processes inside the Sun, the Sun's surface, and its atmosphere that result in solar variability. SDO is also helping scientists to better understand the Sun's influence on Earth and near-Earth space through the use of many wavelengths simultaneously. SDO is also investigating how the Sun's magnetic field is generated and structured.

STEREO (Solar TErrestrial RElations Observatory) was launched by NASA on 26 October 2006. It consists of two nearly identical spacecraft, one orbiting ahead of Earth (A) and the other behind Earth (B). Observations are made simultaneously of the Sun and then combined to provide a 3-D stereo image of the Sun. Spacecraft A takes 347 days to orbit the Sun while spacecraft B takes 387 days. Because the A spacecraft is moving faster than B, they are separating from each other and A is orbiting closer to the Sun than B. The images are adjusted to account for this difference.

Each of the spacecraft carries cameras (a EUV imager and two coronagraphs), particle experiments and radio detectors in four instrument packages. STEREO is used to image the inner and outer corona and the space between Sun and Earth, detect electrons and other energetic particles in the solar wind, study the plasma characteristics of protons, alpha particles and heavy ions, and monitor radio wave disturbances between the Sun and Earth.

By studying the Sun, scientists are gaining a better understanding of how even small solar events effect the operation of technological infrastructure on Earth (such as GPS systems, cable TV, radio and satellite communications).

Features of the Sun

The Sun is a huge ball of burning gas, which contains about 99 % of the mass of the whole solar system. This makes it over 300,000 times as massive as the Earth. The Sun's diameter of 1.4 million km far exceeds Earth's diameter of only 12,760 km. Even the biggest planet—Jupiter, is only one-tenth the diameter of the Sun (see Table 3.2).

The main elements present in the Sun are hydrogen (92 %), followed by helium (7.8 %), and less than 1 % of heavier elements like oxygen, carbon, nitrogen and neon. The Sun is entirely gaseous with an average density 1.4 times that of water. Because the pressure in the core is much greater than at the surface, the core density is eight times that of gold, and the pressure is 250 billion times that on Earth's surface (Table 3.3).

Table 3.2 Details of the Sun

Mass	2.0×10^{30} kg
Relative to Earth	109 times larger
Diameter	1.4 million km
Distance from Earth	150 million km
Density	1410 kg/m^3
Luminosity	3.9×10^{26} J/s
Surface temperature	5500 °C
Interior temperature	15 million degrees Celsius
Equatorial rotation period	25 days
Composition	92 % hydrogen, 7.8 % helium
Surface gravity	290 N/kg (29 × Earth gravity)
Escape velocity	618 km/s
Photosphere thickness	400 km
Chromosphere thickness	2500 km
Core pressure	250 billion atmospheres
Sunspot cycle	11 years
Age	4.5 billion years

Table 3.3 Composition of the Sun

Element	Abundance (percentage of total number of atoms)	Abundance (percentage of total mass)
Hydrogen	91.2	71.0
Helium	8.7	27.1
Oxygen	0.078	0.097
Carbon	0.043	0.40
Nitrogen	0.0088	0.096
Silicon	0.0045	0.099
Magnesium	0.0038	0.076
Neon	0.0035	0.058
Iron	0.0030	0.14
Sulfur	0.0015	0.040

Energy and Luminosity

The Sun produces a 100 million times more energy than all the planets combined. Just over half this energy is in the form of visible light, with the rest being infrared (heat) radiation. Only about a billionth of the Sun's energy reaches us here on Earth.

The Sun's energy comes from the burning of its hydrogen gas via the process of nuclear fusion. In this process four hydrogen atoms combine to make one helium nuclei. During this process some mass is lost and it is this mass that is converted into energy. Every second the Sun converts over 600 million tonnes of hydrogen into helium, and this results in 4.5 million tonnes of matter being converted into energy every second.

Energy generated in the core is carried outward to the surface by radiation and convection processes. Core temperature is about 15 million degrees Celsius, while at the surface the temperature is around 5500 °C. The surface and interior temperature are too hot to have any liquid or solid material.

The luminosity of a star is an indication of the total amount of energy it produces every second. This rate depends on the core temperature and pressure of the star, which in turn depends on its mass. The Sun's luminosity is 3.9×10^{26} J/s.

Throughout its life the Sun has increased its luminosity by about 40 % and it will continue to increase from some time.

Zones of the Sun

The Sun has several different layers or zones of activity. At the centre is the core, which is where energy is produced via nuclear fusion reactions. Above this is the radiative zone, where energy travels very slowly upwards. Closer to the surface is the convective zone where heat is transported much faster to the surface, or photosphere. Surrounding the photosphere is the solar atmosphere that contains two zones—the chromosphere and corona.

The Core of the Sun

The core of the Sun is the central region where nuclear reactions convert hydrogen into helium. These reactions release the energy that ultimately leaves the Sun as visible light. For these reactions to take place a very high temperature is needed. The temperature close to the centre is about 15 million degrees Celsius and the density is about 160 g/cm^3 (i.e. 160 times that of water). Both the temperature and density decrease outwards from the centre of the Sun. The core occupies the innermost 25 % of the Sun's radius. At about 175,000 km from the centre the temperature is only half its central value and the density drops to 20 g/cm^3.

The Radiative Zone

Surrounding the core of the Sun is the radiative zone. This zone occupies 45 % of the solar radius and is the region where energy, in the form of gamma ray photons, is transported outward by the flow of radiation generated in the core. The high-energy gamma ray photons are knocked about continually as they pass through the radiative zone, some are absorbed, some re-emitted and some are returned to the core. It may take the photons a hundred thousand years to find their way through the radiative zone. At the outermost boundary of the radiative zone, the temperature is about 1.5 million degrees, and the density is about 0.2 g/cm^3. This boundary is called the interface layer or tachocline. It is believed that the Sun's magnetic field is generated in this layer. The changes in fluid flow velocities across the layer can stretch magnetic field lines of force and make them stronger. There also appears to be sudden changes in chemical composition across this layer.

The Convective Zone

The outermost zone is called the convective zone, because energy is carried to the surface by a process of convection. It extends from a depth of about 210,000 km up to the visible surface and occupies about 30 % of the Sun's radius. In this zone, plasma gas, heated by the radiative zone beneath, rises in giant convection currents to the surface, spreading out, cooling, and then shrinking—similar to the boiling of water in a pot. Rising cells of gas are visible on the surface as a granular pattern. The granules are around 1000 km in diameter. The convection cells release energy into the Sun's atmosphere. At the surface the temperature is around 5600° and density is practically zero.

   Once the plasma gas reaches the surface of the Sun, it cools and settles back into the Sun to the base of the convection zone, where it receives more heat from the top of the radiative zone. The process then repeats itself. The photons escaping from the Sun, have lost energy on their way up from the core and changed their

Fig. 3.3 Interior structure of the Sun. Energy is transferred by radiation in the inner regions, and by convection in the outer region.

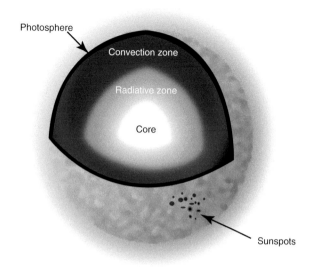

wavelength so most emission is in the visible region of the electromagnetic spectrum.

The lower temperatures in the convective zone allow heavier ions (such as carbon, nitrogen, oxygen, calcium, and iron), to hold onto some of their electrons. This makes the material more opaque so that it is harder for radiation to get through. This traps heat that ultimately makes the fluid unstable and it starts to 'boil' or convect (Fig. 3.3).

The Photosphere

The photosphere is a thin shell of gases about 200 km thick and forms the visible surface of the Sun. Most of the energy radiated by the Sun passes through this layer. It has a temperature of about 5500 °C. From Earth the surface looks smooth, but it is actually turbulent and granular because of convection currents. Material boiled off from the surface of the Sun is carried outward by the solar wind.

The surface of the Sun also contains dark areas called sunspots. Sunspots appear dark because they are cooler than the

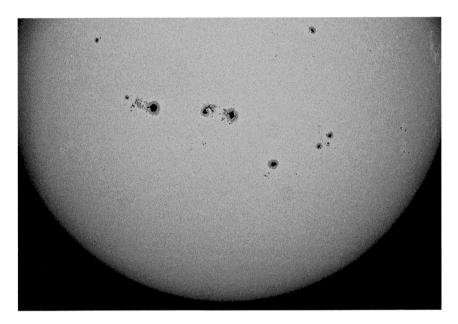

Fig. 3.4 Sunspots on the surface of the Sun. Large sunspots contain *dark umbral centres, grey penumbral haloes,* many large and small single and overlapping spots, and surrounding *whitish plages* (Credit: J. Wilkinson).

surrounding photosphere—about 3500 °C compared to 5500 °C. They radiate only about one fifth as much energy as the rest of the photosphere (see Fig. 3.4).

Sunspots vary in size from 1000 km to over 40,000 km. As they move across the surface of the Sun, sunspots usually change shape—some disappear and new ones appear. Their lifetime seems to depend on their size, with small spots lasting only several hours, while larger spots may persist for weeks or months. The rate of movement of sunspots can be used to estimate the rotational period of the Sun. At the equator, sunspots take about 25 days to move once around the Sun. At the poles sunspots take about 36 days to go around the Sun. This odd behaviour is due to the fact that the Sun is not a solid body like the Earth. Sometimes sunspots appear in isolation, but often they arise in groups.

Sunspots and sunspot groups are directly linked to the Sun's intense magnetic fields. Such spots are areas where concentrated magnetic fields break through the hot gases of the photosphere. These magnetic fields are so strong that convective motion beneath the spots is greatly reduced. This in turn reduces the amount of heat brought to the surface as compared to the surrounding area, so the spot becomes cooler. Data obtained from space probes like SOHO have shown that the strength of the magnetic fields around sunspots is thousands of times stronger than the Earth's magnetic field.

A typical sunspot is about 10,000 km across. Each has two parts: a black central region called the umbra, which in turn is surrounded by a grey region, the penumbra. The darker the area, the lower the temperature. It is possible to view sunspots from Earth by projecting the image of the Sun from a telescope onto a white screen or by using a telescope fitted with a Herschel wedge. Observations of the Sun in this way need to be made carefully so as not to damage the viewer's eyes—NEVER look at the Sun through an unprotected telescope!

Like many other features of the Sun, the number and location of sunspots vary in a cycle of about 11 years. Heinrich Schwabe, a German astronomer, first noted this cycle in 1843. Sunspot maximums occurred in 1968, 1979, 1990 and 2001. Sunspot minimums occurred in 1965, 1976, 1986, 1997 and 2008. The average latitude of sunspots also varies throughout the sunspot cycle. At the beginning of a sunspot cycle, most sunspots are at moderate latitudes, around $28°$ north or south. Sunspots arising much later in each cycle typically form closer to the Sun's equator. The variation in the number of sunspots is now known to be the most visible aspect of a profound oscillation of the Sun's magnetic field that affects other aspects of both the surface and interior (see Figs. 3.5 and 3.6).

The SOHO probe has been able to monitor the Sun for the entire 11-year sunspot cycle (number 23) and the rise of the current cycle (number 24). At the same time, this probe has also monitored the total solar irradiance and variations in the extreme ultraviolet flux, both of which are important to our understanding of the impact of solar variability on Earth's climate (see Fig. 3.5).

Ejection of material from the surface of the Sun often follows solar flares or other solar phenomena. Sometimes this material reaches Earth and gets trapped in the magnetic field around Earth's polar regions. This material consists mostly of charged particles

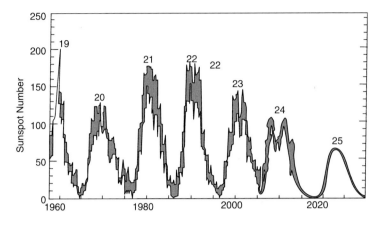

Fig. 3.5 Variations in sunspot numbers tend to go through a maximum/minimum cycle every 11 years. The figure shows sunspot numbers since 1960 with predicted numbers for solar cycles 24 and 25.

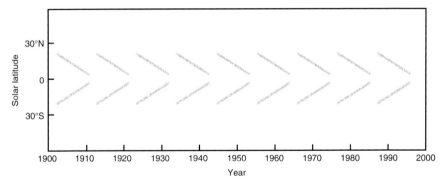

Fig. 3.6 Variations in the average latitude of sunspots.

(ions and electrons), which interfere with communication systems and produce magnetic and ionospheric disturbances such as auroras. An aurora is a bright display of coloured lights in the night sky. Auroras are produced when charged particles (from the Sun) get trapped in the Earth's magnetic field and collide with atoms in our upper atmosphere.

Another visible phenomenon of the Sun's photosphere associated with its magnetic field is the faculae. These are irregular patches or streaks brighter than the surrounding surface. They are clouds of incandescent gas in the upper regions of the photosphere. Such clouds often precede the appearance of sunspots.

The Chromosphere

The chromosphere is the first layer of the Sun's atmosphere. It lies just above the photosphere and is a few thousand kilometres thick. During a solar eclipse, when the Moon passes in front of the Sun, the chromosphere appears as a red shell around the Sun. The chromosphere is much hotter than the photosphere ranging from 4200 °C near the surface to 8200 °C higher up. It consists largely of hydrogen, helium and calcium.

When viewed with a hydrogen alpha filter, dark features called filaments can often be seen against the surface of the Sun. These structures are huge masses of burning plasma ejected upwards from the photosphere and suspended in the Sun's chromosphere and corona by strong magnetic fields. When seen around the limb of the Sun, these eruptions, can be seen as gigantic 'flame-like' structures, and are called **prominences**. The prominences can reach a temperature of 50,000 °C. Some prominences last for only a few hours while others last for weeks. Prominences can only be seen during a total solar eclipse or by using a hydrogen-alpha telescope (Fig. 3.7).

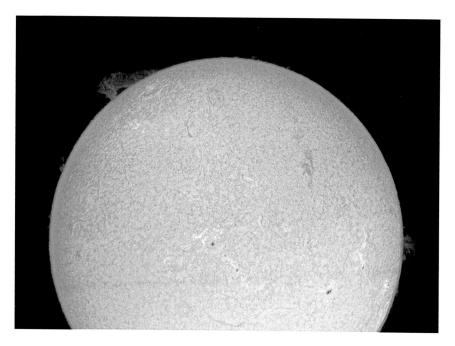

Fig. 3.7 Prominence eruptions on the Sun. Taken by the author through a H-alpha solar telescope on 23rd April 2015 (Credit: J. Wilkinson).

The Corona

The corona is the upper layer of the Sun's atmosphere. During a solar eclipse, it appears as a pale white glowing area around the Sun. Temperatures in the corona reach as high as one million degrees Celsius because of interactions between gases and the photosphere's strong magnetic fields. The corona can extend millions of kilometres into space. The corona consists mainly of ionised gas or plasma.

A **coronal mass ejection** or CME is an expulsion of a part of the corona and ionised particles into space. Such events can represent the loss of several billion tonnes of matter from the Sun at speeds between 10 and 1000 km/s. Some CME's are triggered by solar flares and are associated with strong magnetic fields in the corona. Sometimes, clouds of ejected particles are carried by the solar wind towards Earth (see Fig. 3.8).

A **coronal hole** is a large region in the corona that is less dense and is cooler than its surrounds. They are areas where open magnetic field lines project out from the Sun's surface. Such holes may appear at any time during a solar cycle but they are most common during the declining phase of the cycle. Coronal holes allow denser and faster 'gusts' of the solar wind to escape the Sun. They are sources of many disturbances in Earth's ionosphere and geomagnetic field (see Fig. 3.9).

Solar flares occur when the magnetic field of the Sun changes rapidly to create an explosion of charged particles through the Sun's corona. Such events last from a few minutes to a few hours and can send charged particles, X-rays, ultraviolet rays and radio waves into space. Flares can release energy equivalent to more than a billion one-megaton thermonuclear explosions in a few

Fig. 3.8 In March 2000 an erupting filament lifted off the active solar surface and blasted this enormous bubble of magnetic plasma into space (a coronal mass ejection). The Sun itself (*white circle*) has been blocked out in this picture of the event (Credit: NASA/ESA/SOHO).

seconds. They are sometimes so violent that they cause additional ionisation in the Earth's ionosphere and may disrupt radio communications.

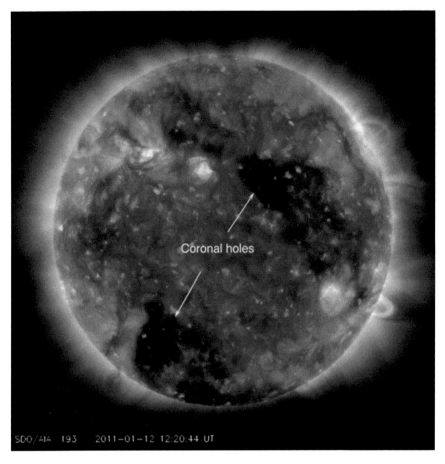

Fig. 3.9 Two coronal holes on the Sun developed over several days in Jan 2011. This UV image was taken by the SDO space probe (Credit: NASA/ SDO).

The Solar Wind

The solar wind is an erratic flow of highly ionised gas particles that are ejected into space from the Sun's upper atmosphere. This wind has large effects on the tails of comets and even has measurable effects on the trajectories of spacecraft. The SOHO, Wind and ACE probes have measured the speed of the solar wind.

The Ulysses space probe provided the first-ever three-dimensional map of the heliosphere from the equator to the poles. Instruments on board the Ulysses space probe also found that the solar wind blows faster around the Sun's poles (750 km/s) than in equatorial regions (350 km/s).

The SOHO space probe found that the solar wind originated from honeycomb-shaped magnetic fields surrounding large bubbling cells near the Sun's poles.

Near the Earth, the particles in the solar wind move at speeds of about 400 km/s. These particles often get trapped in Earth's magnetic field, especially around the poles, and produce auroras.

The solar wind produces a huge bubble in space called the heliosphere. The heliosphere stretches outward from the Sun in all directions to a distance well beyond the world of planets. At its outer boundary, called the heliopause, lies the broader realm of the Oort cloud. The heliosphere is a protective zone, carved out by the solar wind and sustained by the Sun's extended magnetic field. It "protects" in the sense that the solar plasma that flows continually outward from the Sun is strong enough to fend off most of the plasma that comes in stellar winds from other stars, and to keep out all but the most energetic cosmic rays.

Cycles in Solar Activity

While most of the Sun's activity follows the 11-year sunspot cycle, conditions in the heliosphere are driven by a 22-year magnetic cycle. The Sun's magnetic field is like that of a giant bar magnet with a north and South Pole. Data from the Ulysses space probe showed that at during the last solar maximum (2001), the Sun's north and south poles changed places. Ulysses next passed over the Sun's poles during the solar minimum period 2007/2008. At this time the Sun's magnetic polarity was opposite to that of the previous solar minimum.

The Japanese Hinode probe (launched 2006) was the first to be able to measure small changes in the Sun's magnetic field. The magnetic field of the Sun influences the way in which charged particles move through the heliosphere.

Types of Radiation from the Sun

The Sun gives off many kinds of radiation besides visible light and heat. These radiations include radio waves, ultraviolet rays and X-rays. Space probes that orbit the Sun make observations and take pictures in the different wavelengths of electromagnetic radiation.

The Sun's chromosphere and corona are also emitters of radio waves. These were first recorded in 1942 during World War II by British radars as 'radio noise'. Such radio emissions often originate from sunspots and produce what we call 'solar storms'. These radio waves can be collected via radio telescopes on Earth. Observations of the Sun using radio waves provide information different to that obtained from visible wavelengths because the propagation of the two types of radiations are different. For example, coronal gas is transparent to visible light but is opaque to radio waves.

Ultraviolet rays are electromagnetic waves with a shorter wavelength than visible light. They are invisible to the human eye. The Sun gives off more ultraviolet radiation during times of increased solar activity. The Earth's atmosphere absorbs much of this radiation. Scientists have divided the ultraviolet part of the spectrum into three regions: the near ultraviolet, the far ultraviolet, and the extreme ultraviolet. The three regions are distinguished by how energetic the ultraviolet radiation is, and by the "wavelength" of the ultraviolet light, which is related to energy. The near ultraviolet, abbreviated NUV, is closest to optical or visible light. The extreme ultraviolet, abbreviated EUV, is the ultraviolet light closest to X-rays, and is the most energetic of the three types. The far ultraviolet, abbreviated FUV, lies between the near and extreme ultraviolet regions.

X-rays are another form of solar radiation with a very short wavelength. The Sun's X-rays can injure or destroy the tissue of living things. The Earth's atmosphere shields human beings from most of this radiation. Hard X-rays are the highest energy X-rays, while the lower energy X-rays are referred to as soft X-rays. The distinction between hard and soft X-rays is not well defined.

X-rays do not penetrate the Earth's atmosphere. Therefore they must be observed from a platform launched above most of our atmosphere (Fig. 3.10).

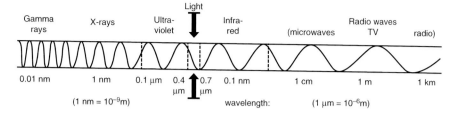

Fig. 3.10 Types of radiations emitted by the Sun.

Solar Eclipses

A **total solar eclipse** is one of the most spectacular astronomical events seen from Earth.

Such an event occurs when the Moon passes directly between the Sun and Earth. An eclipse does not occur at every new moon because the Moon's orbit often passes above or below the Sun instead of directly across it. During such an eclipse, the Moon's shadow traces a curved path across the surface of the Earth. Any person standing in the path of the shadow will see the sky and landscape go dark as the Moon blocks out the sunlight (see Figs. 3.11 and 3.12).

Solar eclipses can be total, partial, or annular, depending on how much Sun is covered by the Moon. Total eclipses occur when the Moon is exactly in line between the Earth and Sun and exactly covers the disc of the Sun. If the Moon is not exactly in line between the Earth and Sun only a partial eclipse occurs. An annular eclipse occurs when the Moon's is far enough away from Earth its apparent size is smaller than the Sun's. Hence a bright ring (annulus) remains visible, and the Sun's corona cannot be seen.

There are as many as two total solar eclipses a year, and sometimes as many as five, but few people have a chance to see them. The paths along which eclipses can be seen are narrow, and totality can last only about seven and a half minutes at most.

Among the features of a total eclipse are the so-called "Bailey's Beads". These are seen just as the Moon's black disc covers the last thin crescent of the Sun. Sunlight shining between the mountains at the Moon's edge looks like sparkling beads. The

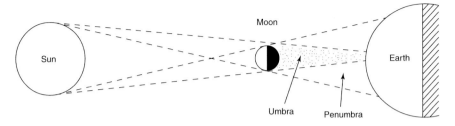

Fig. 3.11 How an eclipse of the Sun occurs.

Fig. 3.12 The solar corona as seen during a total solar eclipse.

Diamond Ring effect is a fleeting flash of light seen immediately preceding or following totality.

At the time of totality, an observer with a small telescope can see the Sun's prominences as long flame-like tongues of incandescent gases around the edge of the Moon's disc. Also during totality, the corona can be seen as a region of glowing gases stretching out from the blacked-out Sun. Care must be taken when observing the Sun, even during an eclipse—and advice should be sought to protect your eyes from damage.

Influence of the Sun on Earth

The Sun has a steady output of charged particles and other matter that is collectively known as the **solar wind**. This wind streams through the Solar System at about 400 km/s. This wind interacts with the atmosphere of Earth and charged particles in particular get trapped in the Earth's magnetic field (the magnetosphere). Our magnetic fields and atmosphere in some ways protect us from some solar radiation and cosmic rays from outer space, but they also let in some of the radiation. Outbursts of radiation from the Sun can have dramatic effects on Earth. When a burst of solar radiation strikes the Earth's magnetic field the result can be geo-magnetic storms that spark huge electric currents and distort the magnetosphere. This can adversely effect radio communication and navigational systems, and pump extra electricity into power lines (sometimes causing blackouts).

Researchers believe that changes in solar activity are having an indirect effect on Earth's climate. Satellite measurements, for example, have detected a small change in the Sun's total output during the course of each sunspot cycle. The ebb and flow of solar radiation can heat and cool the atmosphere of Earth enough to change its circulation patterns, which may have significant impacts on regional weather. Researchers have developed powerful computer models to simulate the impact of the Sun on our climate. One such effort, the Whole Atmosphere Community Climate Model (WACCM), helps researchers study interactions among different levels of the atmosphere, ranging from the surface of Earth to the upper atmosphere and the edge of space. The modelling work is combined with the analyses of data from observing instruments aboard satellites to track the impacts of solar radiation throughout the atmosphere.

Much of what we have learned has been realized in only the last few decades. Solar space missions such as NASA's TRACE (Transition Region and Explorer Spacecraft) and the SOHO (Solar and Heliospheric Observatory) have provided answers to many questions regarding the effect of the Sun on Earth. But there is a lot of work still to be done and many new questions need answering.

The Sun's Future

The Sun is about 4.5 billion years old and is not quite half way through its life cycle. Throughout the second half of its life the Sun is expected to increase gradually in size, luminosity and temperature. In about 5 billion years time the Sun will have expanded to about three times its present size. As the Sun uses up its hydrogen it will become more orange in colour. By this time, temperatures on Earth will be much hotter and all the water will have evaporated. As the Sun continues to expand to about 100 times its present size it will become a red giant engulfing Mercury and Venus. The Earth will be scorched to a cinder. As hydrogen is used up, the core of the Sun will slowly contract, forcing the Sun's central temperature to increase. When the core temperature reaches 100 million degrees Celsius, helium fusion begins to generate carbon and oxygen. Temperature in the core continues to rise causing helium to fuse at an increasing rate. An explosion (the helium flash) will result and a third of the Sun will be blown away. Eventually the Sun will lose its outer layers and contract to become a white dwarf about the size of Earth.

Further Information

See the book "New Eyes on the Sun—a guide to satellite images and amateur observation" by John Wilkinson and published by Springer (2012).
www.solarsystem.nasa.gov (click on the Sun)
www.nineplanets.org (click on the Sun)

4. Mercury: The Iron Planet

Highlights

- NASA's Messenger spacecraft is the first probe ever to orbit the planet Mercury. It began orbiting the innermost planet in the solar system in 2011 and has recorded more than 200,000 photos of Mercury.
- Pictures taken by Messenger of the far side of Mercury show it is a shrinking, ageing planet.
- Mercury is dominated by an iron core, which makes up 85 % of the planet by weight (much larger than previously thought).
- In November 2012, Messenger discovered both water ice and organic compounds in some of the permanently shadowed craters on Mercury's north pole.
- Magnetometers on the Messenger probe found that the source of the magnetic field on Mercury is not dead centre in the planet's interior.

Mercury is the planet in the solar system nearest to the Sun, with an average orbital radius of 58 million km. It is also the smallest planet of the inner solar system with a diameter of only 4880 km, making it about the size of our Moon. Mercury travels fast, taking just 88 days to orbit the Sun, but it takes a slow 58.65 days to rotate once on its axis. Of all the terrestrial planets, Mercury's orbit is the most elliptical. Its elliptical orbit and slow rotation gives it large variations in surface temperatures. During the day temperatures can reach a blistering 430 °C, while at night they can drop to a freezing −180 °C. No other planet experiences such a wide range of temperatures.

Mercury is thought to have formed at the same time as the other planets in the solar system about 4.5 billion years ago. Because it is so close to the Sun, Mercury must have been very hot and in a molten state before it cooled to become a solid planet.

J. Wilkinson, *The Solar System in Close-Up*, Astronomers' Universe, DOI 10.1007/978-3-319-27629-8_4,

Fig. 4.1 Mercury as seen by the Messenger probe during its 2008 flyby of the planet. The sprawling Caloris basin (*upper right*) is one of the solar system's largest impact basins. Created during the early history of the solar system by the impact of a large asteroid-sized body, the basin spans about 1500 km and is seen in *yellowish hues* in this enhanced colour mosaic. *Orange splotches* around the basin's perimeter are now thought to be volcanic vents, new evidence that Mercury's smooth plains are indeed lava flows (Credit: NASA).

As Mercury cooled, it also began to contract. Since it formed, the surface of Mercury has been churned up by many meteorite impacts (Fig. 4.1).

Early Views About Mercury

Mercury has been known since the time of the Sumerians (3rd millennium BC). The planet was given two names by the ancient Greeks: Apollo for its apparition as a morning star and Hermes as an evening star. Greek astronomers knew, however, that the two names referred to the same body. To the Greek's, Hermes was the messenger of the Gods. In Roman mythology Mercury was the god of commerce, travel and thievery. The planet probably received this name because it moves so quickly across the sky.

Giovanni Schiaparelli using a simple telescope made the first map of Mercury in the 1880s. The map only showed areas of dark and light. A more detailed map was produced by Eugenios Antoniadi between 1924 and 1933, but has since been proved inaccurate. Both these astronomers believed Mercury to rotate once on its axis in 88 Earth days, with one hemisphere permanently facing the Sun. This meant that Mercury's day was the same length as its year. However, radar measurements carried out in the early 1960s showed that the true axial rotation period was 58.6 days. Thus it is now known that Mercury rotates three times during two orbits of the Sun. The result of this is that the same hemisphere is pointed towards Earth every time the planet is best placed for observation. This effect also means that the Mercurian day (sunrise to sunset) is 176 Earth-days long, or two Mercurian years.

Antoniadi also believed that Mercury had an atmosphere because he thought he could see clouds above its surface. We now know that Mercury's atmosphere is far too tenuous to support clouds. The lack of clouds is also due to the fact that Mercury's escape velocity is only 4.3 km/s, so any gas particles would be moving too quickly to be restrained by Mercury's gravity (Table 4.1).

Table 4.1 Details of Mercury

Distance from Sun	57,910,000 km (0.38 AU)
Diameter	4880 km
Mass	3.3×10^{23} kg (0.055 Earth's mass)
Density	5.43 g/cm^3 or 5430 kg/m^3
Orbital eccentricity	0.206
Period of revolution (length of year)	88 Earth days or 0.241 Earth years
Rotation period	58.65 Earth days
Orbital velocity	172,400 km/h
Tilt of axis	2°
Day temperature	430 °C
Night temperature	−180 °C
Number of Moons	0
Atmosphere	Practically none (some oxygen, sodium, helium)
Strength of gravity	3.3 N/kg at surface

Probing Mercury

Mercury is the least explored of our solar system's inner planets. To date the planet has been visited by only two spacecraft—Mariner 10 and Messenger. Mariner 10 flew past Venus on 5 February 1974, in order to get a gravity assist to Mercury. It flew by the planet three times between March 1974 and March 1975. Mariner 10 was also the first spacecraft to have an imaging system, and the encounter produced over 10,000 pictures that covered 57 % of the planet. Mercury is too close to the Sun to be mapped by the Hubble Space Telescope.

Another NASA spacecraft, called **Messenger**, was launched on a mission to Mercury on 2 August 2004. Messenger stands for 'MErcury Surface, Space ENvironment, GEochemistry and Ranging'. This probes 7-year journey included 15 trips around the Sun, one Earth fly-by, two Venus fly-bys and three Mercury fly-bys (January 2008, October 2008, September 2009) before it entered orbit around Mercury in March 2011. The fly-bys helped focus the science mission before the spacecraft entered orbit. With a package of seven scientific instruments Messenger has been able to determine Mercury's composition, map its surface and magnetic field, measure the properties of its core, explore the mysterious polar deposits to learn whether ice lurks in permanently shadowed regions, and characterise Mercury's tenuous atmosphere and Earth-like magnetosphere.

Pictures taken by Messenger in January 2008 show the far side of Mercury contains wrinkles of a shrinking, ageing planet. There are scars from volcanic eruptions and craters with a series of troughs radiating from them (Figs. 4.2 and 4.3).

Messenger's primary mission finished on 17 March 2012, having collected over 100,000 images. In November 2013, NASA managers decided to extend the mission through to March 2015.

Fig. 4.2 Messenger's looping polar orbit around Mercury ranges in altitude from just 200 km to about 15,000 km. The spacecraft uses the close-ins for observations and beams back its data to Earth when it is far out.

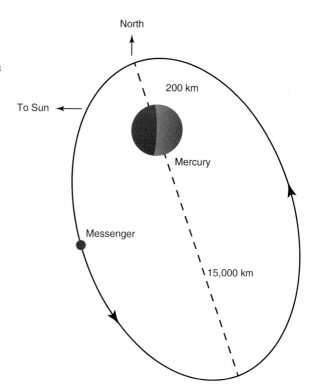

Messengers mission finally ended on 30 April 2015 when it crashed into the surface of Mercury creating a new crater. The mission exceeded all expectations.

A joint European-Japanese mission, to Mercury is due for launch on 15 August 2015. The spacecraft, called BepiColombo, will arrive at Mercury in January 2022 for a 1-year nominal mission with a possible 1-year extension. BepiColombo will use the gravity of the Earth, Venus and Mercury in combination with solar-electric propulsion to journey to Mercury. When approaching Mercury, the spacecraft will use the planet's gravity plus conventional rocket engines to insert itself into a polar orbit (Table 4.2).

Fig. 4.3 The Messenger space probe took this photo of the far side of Mercury in January 2008. The previously unseen features suggest the planet is old and wrinkly (Credit: JHU/NASA).

Table 4.2 Significant space probes sent to Mercury

Name of probe	Country of origin	Date launched	Notes
Mariner 10	USA	1973	Now in solar orbit
Messenger	USA	2004	Orbiting Mercury 2011–2015

Position and Orbit

Mercury has the most eccentric orbit of all the planets. At closest approach (perihelion) it is only 46 million km from the Sun but its furthest distance (aphelion) is 70 million km. At perihelion Mercury travels around the Sun at a very slow rate.

Observation of Mercury is difficult because of its close proximity to the Sun. The best time to view the planet is twice a year when it appears above the horizon at its greatest distance from the Sun. At these times, Mercury can be seen just before sunrise or just after sunset as an orange object.

The axis of rotation of Mercury is almost vertical. This means the plane of its equator coincides with the plane of its orbit.

Mercury has the shortest year of any planet, taking only 88 days to orbit the Sun. It has no known satellites (moons). Because it is closer to the Sun than Earth, Mercury is seen to go through phases just like our Moon. Mercury's size appears to vary according to its phases because of its changing distance from Earth. When Mercury first appears in the evening sky, it is coming around the far side of its orbit toward us, and through a telescope appears as a full crescent.

At rare intervals, observers from Earth can see Mercury pass in front of the Sun. Such a passing is called a transit. Transits occurred on 7 May 2003 and 8 November 2006. A transit should be viewed by projecting the Sun's image from a telescope, onto a white screen. A planet would appear as a black dot slowly moving across the Sun's image. Care should always be taken when viewing the Sun—never look directly at the Sun with a telescope.

Density and Composition

Mercury is thought to be one of the densest planets in the Solar System. This high density suggests that an iron core that makes up 85 % of the planet by weight dominates Mercury. During 2007, a team of astronomers announced that they have evidence suggesting that some of the core is molten. Bouncing radio waves off the planet and analysing the return signals made the discovery. The latest information suggests the solid iron core is surrounded by a liquid outer layer of iron, followed by a layer of iron-sulfide and a thin, solid silicate mantle and crust (see Fig. 4.4). Little is known about Mercury's crust, but it is thought to extend down less than 100 km. The crust seems to have cooled rapidly once formed, and is solid enough to preserve surface features.

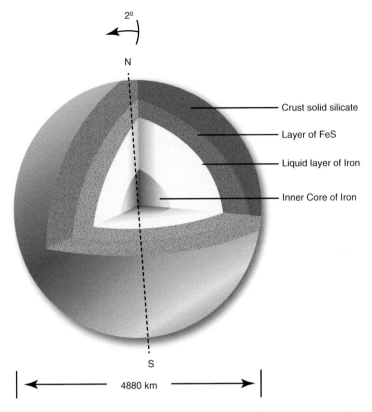

Fig. 4.4 The interior of Mercury as determined by Messenger data. The core contains mostly iron and is larger than previously thought (85 % radius).

The Messenger probe has found sodium, magnesium, calcium, oxygen, helium and even water molecules streaming away from the planet. These streams are not uniform, with sodium and calcium concentrated over the polar regions, and magnesium and helium more evenly distributed. Bursts of energetic electrons are regularly detected whenever the Messenger probe passes over the mid-northern latitudes of Mercury (the source of these electrons is unknown).

Mercury is about one third the size of Earth and a little more than one-third the gravity of Earth. A 75 kg person on Earth has a weight of 735 N, but on Mercury the same person would weigh 247 N.

The Surface

The surface of Mercury is heavily cratered and looks very similar to our Moon. Most of the craters are impact craters formed from bodies colliding with the surface. The distribution of craters on Mercury, the Moon and Mars are similar, suggesting the same family of objects was responsible for the impacts on each body. One of the largest impact features on Mercury is the Caloris Basin. This basin measures about 1300 km across and has been partially flooded with lava from volcanic activity. It was probably formed by a very large impact early in the history of the Solar System. As the basin floor settled under the weight of volcanic material, fractures and ridges formed (Figs. 4.5 and 4.6).

In addition to the heavily cratered areas, Mercury also has large regions of smooth plains caused by ancient lava flows. Mariner 10 also revealed some large escarpments over the surface, some up to hundreds of kilometres in length and 3 km high. Some of these cut through the rings of craters and others seem to be formed

Fig. 4.5 Cameras on the Messenger probe took this image of lava flooded craters and large areas of smooth volcanic plains on Mercury (Credit: NASA/JHU).

Fig. 4.6 The southern hemisphere of Mercury as seen by cameras on the Messenger probe in April, 2013. This image highlights Bach crater (*lower double ringed*) and the bright rays of Han Kan crater (Credit: NASA/JHU).

by tectonic forces. The largest scarp or cliff observed to date is Discovery Rupes, which is about 500 km long and 3 km high. Such scarps are thought to be due to global compression and tectonic activity as Mercury cooled.

A reanalysis of data collected by Mariner 10, suggests that recent volcanism has occurred on Mercury. Many of Mercury's craters are not covered by volcanic flows indicating that they were formed after volcanic flows ceased. Heavy bombardment of the planet ended about 3.8 million years ago.

Recent radar analysis has found that a number of craters near each pole have high radar reflectivity, suggesting the presence of ice. The interiors of these polar craters are permanently shaded from the Sun's heat, making the preservation of ice possible. In November 2012, NASA announced that Messenger had discovered both water ice and organic compounds in some of the permanently shadowed craters on Mercury's north pole.

Data from cameras on the Messenger probe have confirmed that broad lava plains fill most of the land between the planet's abundant craters. The north pole is distinctly lower in elevation than elsewhere, and was probably inundated early in Mercurian history by a volcanic outpouring of almost unimaginable size. Although there is no evidence of recent or ongoing eruptions, Mercury is not geologically dead. Its surface abounds with clusters of small shallow pits, ranging in size from tens of metres to several kilometres across. These hollows appear to be associated with some unknown black material found scattered over the surface.

Instruments on the Messenger probe have also shown that very little iron exists on Mercury's surface. Instead, the rocks are infused with lots of magnesium, a chemical signature unique among the terrestrial worlds. There are also high abundances of sulfur, potassium, and sodium. These volatile elements vaporise at relatively low temperatures and their presence has caused scientists to rethink their ideas on how the planet formed.

In February 2013, NASA published the most detailed and accurate 3D map of Mercury to date, assembled from thousands of images taken by Messenger.

Mercury's Atmosphere

Mercury has a very thin atmosphere. In 1974 the Mariner 10 space probe detected traces of oxygen, sodium, helium, potassium and hydrogen vapours. Earth based telescopes have detected gaseous sodium, potassium and calcium. In 2008, the Messenger spacecraft discovered magnesium and water vapour in the atmosphere of Mercury. The hydrogen and helium may have originated from the solar wind while the sodium may have come from surface rocks bombarded by the wind or meteorites. Astronomers have observed clouds of sodium vapour occasionally rising from the surface of Mercury. There are striking differences in the amounts of calcium, magnesium and sodium when the planet was closer to and further from the Sun.

The atmospheric pressure is only a million-billionth that of Earth's—as low as many vacuums created in Earth laboratories. Mercury has very little atmosphere because its gravity is too weak to retain any significant gas particles, and it is so hot that gases quickly escape into outer space. The quantities of these gases have been poorly determined and vary depending on the position of Mercury in its orbit. The atmospheric gases are much denser on the cold night-side of Mercury than on the hot dayside.

Temperature and Seasons

Because of its closeness to the Sun, surface temperature variations are extreme on Mercury—more than on any other planet. You could roast during the day at 430 °C, and freeze during night at −180 °C. The high day temperature would be hot enough to melt the metals zinc and tin. One day on Mercury is equal to 58.65 Earth days, so it takes a long time to warm up from the cool of night and it takes a long time to cool down from the heat of day.

Mercury is not the hottest planet on average. The temperature on Venus is slightly hotter than on Mercury but is more stable because of thick clouds on that planet.

On Earth, the seasons change in a regular way due because the rotational axis is inclined (at 23.5° from the perpendicular to its orbital plane). As a result, each hemisphere on Earth receives more direct sunlight during one part of the orbit than the other.

Mercury's axis is very near perpendicular to its orbital plane (2° from perpendicular), so no seasonal changes occur. Some craters near Mercury's poles never receive any sunlight and are permanently cold (Figs. 4.7 and 4.8).

Fig. 4.7 The International Astronomical Union (IAU) has named an impact crater on the planet Mercury after John Lennon, the British pop music sensation who helped make The Beatles the most popular group of their generation. Lennon is one of ten newly named craters on the planet, joining 114 other craters named since NASA's Messenger spacecraft's first Mercury flyby in January 2008 (Credit: NASA/Johns Hopkins Applied Physics Lab/Carnegie Institution).

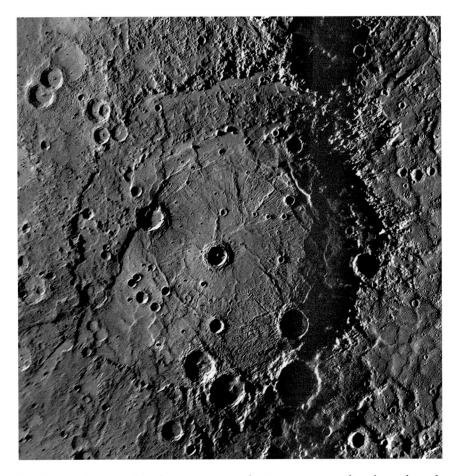

Fig. 4.8 An image taken by cameras on the Messenger probe of Rembrandt impact crater on Mercury. The surface contains impact craters, wrinkle ridges and plains covered with lava (Credit: NASA/JHU).

Magnetic Field

The magnetic field of a planet is generated by electric currents flowing in its molten metal core as the planet rotates. Because Mercury rotates slowly (59 times slower than Earth) it was not expected to have a magnetic field. In 1974 Mariner 10 did detect a very weak magnetic field but the Messenger probe was able to measure it more accurately. The magnetic field tends to be stronger at the equator than at other areas of Mercury but is only 1 % the intensity of Earth's magnetic field. Magnetometers on the

Messenger found that the source of the magnetic field is not dead centre in the planet's interior; rather it is offset toward the north pole by 480 km, about 20 % of Mercury's radius. This northward offset leaves the planet's southern hemisphere more vulnerable to bombardment by space radiation. The offset also suggests that the dynamo responsible for the magnetic field originates not in the planet's heart but closer to its core-mantle boundary.

The Messenger spacecraft also found that Mercury's magnetic field is responsible for several magnetic 'tornadoes'—twisted bundles of magnetic fields connecting the planetary field to inter-planetary space—these are around 800 km wide or a third the total radius of the planet.

Further Information

www.nasa.gov/mission_pages/messenger/main/index.html
www.space.com/36-mercury-the-suns-closest-planetary-neighbor. html
http://solarsystem.nasa.gov (and click on Mercury)

5. Venus: A Hot, Toxic Planet

Highlights

- Venus was volcanically active as recently as 2.5 million years ago.
- Venus is a much more inhospitable world than Earth, with surface temperatures topping 450 °C and a super-dense atmosphere composed of toxic gases.
- Venus has an unusual super-rotating upper atmosphere, which flies around the planet once every 4 days.
- A huge double atmospheric vortex exists at the south pole of Venus.
- Flashes of lightning regularly occur in the sulfuric acid clouds of Venus.
- In July 2014 the Venus Express probe used aerobraking maneuvers to lower its orbit to within 250 km of Venus's north pole (just above the top of the atmosphere). It found conditions to be more variable than previously thought.

Venus is the second planet from the Sun, orbiting on average at a distance of 108 million km from the Sun. It is the sixth largest planet with a diameter of 12,104 km. Venus is sometimes regarded as Earth's sister planet since it is similar in size and mass to Earth. Venus orbits the Sun between Mercury and Earth but is twice as far from the Sun as Mercury. It comes closer to Earth than any other planet in the solar system.

One of the strange things about Venus is that it spins on its axis in the opposite direction to that of the other terrestrial planets—it seems to be upside down with its north and south poles reversed. The unusual state is thought to be due to a massive impact early in the planet's life.

Venus is thought to have formed at the same time as the other planets in the Solar System about 4.5 billion years ago. Because it

J. Wilkinson, *The Solar System in Close-Up*, Astronomers' Universe,
DOI 10.1007/978-3-319-27629-8_5,
© Springer International Publishing Switzerland 2016

is close to the Sun, Venus must have been hot and in a molten state before it cooled to become a solid planet. Out of all the planets, Venus still has the hottest surface temperature, even though it is not the closest planet to the Sun. This is mainly due to its dense atmosphere, which traps heat and pushes down on the surface with a pressure 92 times that experienced on Earth. In fact temperatures are hot enough to melt the metals tin, zinc and lead.

The surface of Venus is completely covered by thick clouds and this makes it impossible to see the surface from Earth or from space without special radar imaging techniques.

Venus has no natural satellites (Moons) (Fig. 5.1).

Early Views About Venus

Venus is the brightest planet as seen from Earth. At certain times of the year it can be seen in the evening sky just after sunset; at other times of the year it appears to rise in the east just before sunrise.

Venus was well known to the ancient Greeks and Romans because of its brightness in the night sky, but the Greeks believed Venus to be two different objects: Phosphorus as the morning star and Hesperos as the evening star.

To the ancient Romans, Venus was the goddess of love and beauty (Venus is the only planet named after a goddess). The brightness of Venus as seen from Earth is due to its covering of dense clouds which reflect over three-quarters of the sunlight received by the planet. These clouds completely hide the surface of the planet from view (Table 5.1).

Probing Venus

People have long thought Venus being close to Earth and similar in size and mass, would have conditions suitable for life. However, early space probes sent to Venus disproved this theory. The probes found Venus to have a hostile-to-life or hell-like environment. The

Fig. 5.1 The planet Venus is completely covered by dense clouds. In order to see the surface, special radar imaging techniques need to be used. This image was taken by the Messenger probe during a recent flyby of the planet (Credit: NASA).

thick atmosphere, high surface temperature and high pressure hampered early exploration by spacecraft, and many probes were unsuccessful. Both the USSR and USA have sent more probes to Venus than any other planet, mainly because of its closeness to Earth.

During the 1960s the USSR launched a series of Venera space-craft on missions to Venus. Venera 1 was the first space probe to flyby Venus in 1961, but communications with Venera 2 and 3 failed just before arrival. The first successful probe to enter the Venusian atmosphere was Venera 4 on 18 October 1967. Although

Table 5.1 Details of Venus

Distance from Sun	108,200,000 km (0.72 AU)
Diameter	12,104 km
Mass	4.87×10^{24} kg (0.82 Earth's mass)
Density	5.25 g/cm^3 or 5250 kg/m^3
Orbital eccentricity	0.007
Period of revolution	224.7 Earth days
Rotation period	243 Earth days
Length of year	0.615 Earth years
Orbital velocity	126,108 km/h
Tilt of axis	177.3°
Day temperature	480 °C
Night temperature	470 °C
Number of Moons	0
Atmosphere	Carbon dioxide
Strength of gravity	8.1 N/kg at surface

this craft was crushed during descent, it sent back useful data on the planet's atmosphere including its chemical composition, pressure, and temperature. In 1969, both Venera 5 and 6 returned data indicating an atmosphere of 93–97 % carbon dioxide. Venera 6 returned data down to within 26 km of the surface before being crushed by the pressure. Venera 7 achieved the first successful landing of a spacecraft on any planet on 15 December 1970. The probe used an external cooling device to allow it to send back 23 min of data. Venera 9 included an orbiter and a lander—the lander arrived on the Venusian surface on 22 November 1975 and transmitted the first black and white images of the planets surface. Venera 13 survived for 2 h and 7 min on the Venusian surface. It took colour images and analysed a soil sample. The first colour panoramic views of the surface were sent back by Venera 14 in November 1981. This probe also conducted soil analysis using an X-ray fluorescence spectrometer. Venera 15 and 16 were the first spacecraft to obtain radar images of the surface from orbit. The images were used to produce a map of the northern hemisphere from the pole to 30° north latitude. During 1985 Vega 1 and 2 flew by Venus on their way for a flyby of comet Halley. Vega 1 dropped off a Venera-style lander and a balloon to investigate the cloud layers. The lander from Vega 1 failed, but Vega 2's lander was able to collect soil samples. Both Vega 1 and 2 are now in solar orbit.

The USA through NASA was also active in sending probes to Venus. The first USA probe to make a flyby of Venus was Mariner

2 on 14 December 1962. Mariner 2 passed Venus at a distance of 34,800 km and scanned its surface with infrared and microwave radiometers, showing the surface temperature to be about 425 °C. Mariner 5 confirmed the temperatures in 1967 as it passed within 3900 km of the planet. It also studied the magnetic field and found an atmosphere containing 85–97 % carbon dioxide. Mariner 10 flew past Venus on 5 February 1974 for a gravity assist to the planet Mercury. It recorded circulation in the atmosphere of Venus and showed the temperature of the cloud tops to be −23 °C.

In 1978 the USA launched two Pioneer Venus probes to orbit Venus. Pioneer Venus 1 (also known as Pioneer 12) operated continuously from 1978 until 8 October 1992, when contact was lost and it burnt up in the Venusian atmosphere. The orbiter was the first probe to use radar imaging in mapping the planet's surface. Pioneer Venus 2 (also known as Pioneer 13) carried four atmospheric probes that were released on 9 December 1978. The four probes descended by parachute and collected data on the atmospheric layers before burning up in the atmosphere. One of the sub-probes landed intact and sent back data for over an hour.

The Galileo spacecraft flew past Venus on its way to Jupiter in February 1990. The USA also sent the Magellan spacecraft into orbit around Venus in August 1990. This probe was launched from the space shuttle Atlantis in May 1989 and took 15 months to reach Venus. Its main mission was to produce a high-resolution map of Venus using synthetic aperture radar, which can see through clouds. The spacecraft mapped 99 % of the planet's surface using a polar orbit. In 1994, the craft was directed into the atmosphere where it burned up (Fig. 5.2).

The **Venus Express** probe was launched by the European Space Agency on 9 November 2005. On 11 April 2006 the probe went into polar orbit around Venus. At closest approach, the probe was about 250 km above the north pole and 66,000 km above the south pole. The main objectives of the mission include exploring the global circulation of the Venusian atmosphere, chemistry of the atmosphere, surface volcanism and atmospheric loss. Thermal imaging done by the probe showed a thick layer of clouds, located at about 60 km altitude that traps heat radiating from the surface. In December 2014 the probe ran out of fuel and contact was lost. The spacecraft was expected to plunge into the atmosphere of

Fig. 5.2 The Magellan spacecraft showed Venus is covered with extensive lava flows and lava plains. This global view of the surface of Venus was produced by the Solar System Visualization project and the Magellan science team at the JPL Multimission Image Processing Laboratory. The *bright areas* are the equatorial highlands known as Aphrodite Terra (Credit: NASA).

Venus in January 2015, falling in pieces, corroding and melting, toward the searing planetary surface.

On 24 October 2006 NASA's Messenger probe (while en route to Mercury) made a flyby of Venus at an altitude of 3000 km. During the encounter, Messenger passed behind Venus and entered superior conjunction, a period when Earth was on the exact opposite side of the Solar System, with the Sun inhibiting radio contact. For this reason, no scientific observations were

Table 5.2 Recent space probes to Venus

Spacecraft	Country of origin	Date launched	Notes
Vega 1	USA	1984	Flyby with lander
Vega 2	USA	1984	Flyby with lander
Galileo	USA	1989	Flyby
Magellan	USA	1989	Radar mapping 99 %
Messenger	USA	2004	Flyby
Venus Express	Europe	2005	Polar orbiter
Akutsuki	Japan	2010	Flyby; retry 2015
IKAROS	Japan	2010	Flyby, solar sail test
Shin'en	Japan	2010	Flyby, failed

conducted during the flyby. Communication with the spacecraft was reestablished in late November and it performed a deep space maneuver on 12 December, to correct the trajectory to encounter Venus in a second flyby. On 5 June 2007, Messenger performed a second flyby of Venus at an altitude of 338 km, for the greatest velocity reduction of the mission. The encounter provided visible and near-infrared imaging data of the upper atmosphere of Venus. Ultraviolet and X-ray spectrometry of the upper atmosphere were also recorded, to characterize the composition. The ESA's Venus Express was also orbiting during the encounter, providing the first opportunity for simultaneous measurement of particle-and-field characteristics of the planet (Table 5.2).

Position and Orbit

Venus orbits the Sun in a nearly circular orbit as shown by its small orbital eccentricity. Its mean distance from the Sun is just over 108 million km and it passes within 40 million km of Earth, closer than any other planet.

Observation of Venus is easy because of its close proximity to Earth, and it is the brightest object in the sky (apart from the Sun and Moon). Venus can be seen either in the eastern sky before sunrise or in the western sky after sunset. It is so bright is can often be seen during daylight and is often mistaken for an Unidentified Flying Object (UFO).

Venus orbits the Sun in about 225 Earth days. Because it is closer to the Sun than Earth, Venus is seen to go through phases just like our Moon. Venus's size appears to vary according to its phases because of its changing distance from Earth. Venus is in full phase when furthest from Earth, and when close to Earth it is seen as a thin crescent phase. It is possible to view the phases through binoculars or a small telescope.

As Venus and Earth travel around the Sun, Venus can be seen near the opposite side of the Sun about every 584 days. When Venus is moving toward Earth, the planet can be seen in the early evening sky (west). When Venus is moving away from Earth, the planet can be seen in the early morning sky (east).

At rare intervals, observers from Earth can see Venus transit or pass in front of, the Sun. A last transit occurred on 8 June 2004, and again on 6 June 2012. A transit should be viewed by projecting the Sun's image from a telescope, onto a white screen. The planet can be seen as a black dot slowly moving across the Sun's image. Care should always be taken when viewing the Sun—never look through a telescope at the Sun with your eyes.

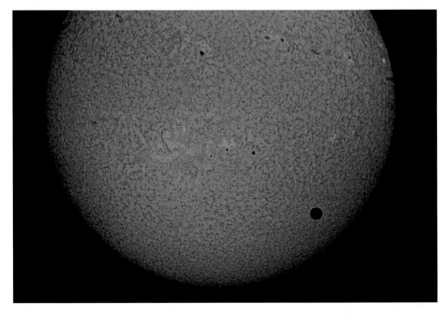

Fig. 5.3 The 2006 transit of Venus across the Sun as recorded by the author through a H-alpha solar telescope (Credit: J. Wilkinson).

Density and Composition

Venus is a rocky planet, much like the Earth. Given its similar size, mass, and density to our planet, scientists think that its interior is much like Earth's own. In addition to a crust significantly older than Earth's constantly changing surface, Venus probably also has a mantle and a core.

Venus is the third densest planet in the solar system. This high density is due to the fact that Venus probably has a large rocky core made of mostly nickel and iron. The core is about 3340 km in radius and is surrounded by a molten silicate mantle about 2680 km thick. There is also a thin outer layer or crust about 50 km thick similar to the crust on Earth. Recent data from the Magellan probe indicates the crust is stronger and thicker than had previously been assumed. Future spacecraft will deposit seismometers to search for 'earthquakes' that can help scientists probe the planet's interior (see Fig. 5.4).

The strength of gravity on Venus is slightly less than that on Earth. A 75 kg person on Earth would weigh 735 N, but on Venus they would only weigh 607 N.

The Surface

The surface of Venus cannot be seen from Earth because of the thick clouds surrounding the planet. However, some features have been detected by radar.

In the 1960s, both the USSR and USA began sending space probes to Venus. Reaching the surface proved to be more difficult than anyone thought because the atmospheric pressure was so great that many early craft were crushed.

Instruments on Venera 7 found the surface temperature was 475 °C and the surface pressure 90 atm (about the same as the pressure at a depth of 1 km in Earth's oceans). The black-and-white photographs returned from the Venera 9 lander showed a rocky terrain with basaltic stones several centimetres across and soil scattered between them. The temperature at the landing site was 460 °C and wind speed was only 2.5 km/h. The terrain around the

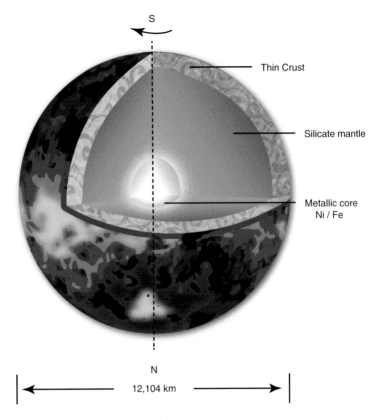

Fig. 5.4 The interior structure of Venus.

lander from Venera 10 was more eroded than at the Venera 9 landing site.

Soil analysis by Venera 14 showed the surface rock type to be basaltic, similar to that found at mid-ocean ridges on Earth.

Venera 15 and 16 produced a map of the northern hemisphere from the pole to 30°N and found several hot spots that possibly were caused by volcanic activity. Most of the surface of Venus consists of gently rolling plains with no abrupt changes in topography. There are also several broad depressions called lowlands, and some large highland areas including two named Aphrodite Terra and Ishar Terra. The highland areas can be compared to Earth's continents, and the lowland areas to its ocean basins. Several canyons and a few rift valleys were also mapped.

There are no small craters on Venus, probably because small meteoroids burn up in Venus' dense atmosphere before reaching the surface. There are some large craters and these appear to come in bunches suggesting that large meteoroids break up into pieces just before hitting the ground.

The best pictures of the Venusian surface came from the orbiting US spacecraft Magellan, which produced detailed maps of Venus' surface using radar. Images from Magellan shows that much of the surface is covered by flat rolling plains with several highland regions. The highest mountain on Venus is Maxwell Montes, situated near the centre of Ishtar Terra and rising to around 11 km above the mean surface level. Maxwell Montes is about 870 km long.

Infrared measurements by Venus Express suggest that Venus might have had a system of plate tectonics in the past, as Earth does today, as well as an ocean of liquid water. Other observations indicate that the planet was likely volcanically active as recently as 2.5 million years ago. Venus may still be active volcanically, but only in a few hot spots. There is also evidence of lava flows, volcanic domes, collapsed volcanic craters and volcanic plains. There are also several large shield type volcanoes (similar to those at Hawaii). Two large and possibly active shield volcanoes are Rhea Mons and Theia Mons, which tower 4 km high.

Planetary scientists believe Venus loses heat from its interior via hot spot volcanism rather than via convection as in the case of Mercury. Hot spots produce shield volcanoes and flood volcanism and these features are common on the present day Venusian landscape. The Magellan probe also revealed surface features called arachnoids that look like craters with spider legs radiating from them. These features are thought to have formed when molten magma pushes up from the interior with such force that the surrounding crust gets cracked (Fig. 5.5).

Tectonic movement that has resulted in crustal shortening, stacking of crustal blocks and wrinkle ridges on the lowlands and rolling plains may have also shaped the surface of Venus. Also seen on the lowlands and plains are fractures formed when the crust was stretched or pulled apart. Diana Chasma is the deepest

Fig. 5.5 False-colour image of the Venusian volcano Sapas Mons as pro-duced by Magellan. Sapas Mons is approximately 400 km across and 1.5 km high (Credit: NASA).

fracture on Venus with a depth around 2 km below the surface and a width of nearly 300 km.

The oldest terrains on Venus are about 800 million years old. Lava flows at that time probably wiped out earlier surface features and larger craters from early in Venus' history.

Approximately 300–500 million years ago there was massive resurfacing on Venus which may have "turned off" any plate tectonics on the planet, completely solidifying the crust into a single surface.

Venus probably once had large amounts of water but the high temperature boiled all this away, so Venus is now very dry (Fig. 5.6).

Fig. 5.6 Aine Corona is the large circular structure near the center of this Magellan radar image. It is approximately 200 km in diameter. Just north of Aine Corona is one of the flat-topped volcanic constructs known as 'pancake' domes for their shape and flap-jack appearance. This pancake dome is about 35 km in diameter and is thought to have formed by the eruption of extremely viscous lava. Complex fracture patterns like the one in the *upper right* of the image are often observed in association with coronae and various volcanic features (Credit: NASA).

The Atmosphere

Many of the probes sent to Venus have provided information about its atmosphere. The clouds of Venus conceal a hostile atmosphere that reaches a height of about 250 km. Most of the atmosphere however is concentrated within 30 km of the surface.

The first probe to be placed directly into the atmosphere and to return data was Venera 4 in 1967. It found that the atmosphere was 90–95 % carbon dioxide with clouds of sulfuric acid droplets. Mariner 5 arrived at Venus 1 day after Venera 4 and passed within 3900 km of the planet's surface—it also found an atmosphere dominant in carbon dioxide. Venera 5 and 6 reported an atmosphere of 93–97 % carbon dioxide, 2–5 % nitrogen, and less than 4 % oxygen. These two probes returned data to within 26 km and 11 km respectively of the surface before being crushed by the high atmospheric pressure.

From 1978 to 1988 the amount of sulfur dioxide in the atmosphere decreased by 10 %. The reason for this decrease may have been due to a decrease in volcanic activity during this period.

In 1978, the Pioneer Venus 2 probe detected a fine haze in the atmosphere at a height of 70–90 km. Between 10 and 50 km there was some atmospheric convection and below 30 km the atmosphere was clear.

The Venus Express probe found Venus has an unusual super-rotating upper atmosphere, which flies around the planet once every 4 days, in stark contrast to the rotation of the planet itself, 243 days. By tracking the movements of distinct features in the cloud tops over the last 6 years, scientists have been able to monitor patterns in the long-term global wind speeds. Venus Express determined that wind speeds have mysteriously increased from 300 to 400 km/h over a span of 6 years. The probe also discovered a huge double atmospheric vortex at the south pole of the planet, and flashes of lightning that regularly occur in Venus's sulfuric acid clouds. In 2011 a layer of ozone was detected in the upper atmosphere of Venus (see Fig. 5.7).

Unlike the clouds on Earth, which appear white from above, the cloud tops on Venus appear yellowish or yellow–orange. These colours are thought to be due to sulfur and sulfur compounds in the atmosphere. Evidence suggests these compounds have originated from volcanic activity.

In July 2014 Venus Express used aerobraking maneuvers to lower its orbit to within 250 km of Venus's north pole (just above the top of the atmosphere). The results show that the atmosphere seems to be more variable than previously thought for this altitude range.

Between altitudes of 165 and 130 km, the atmospheric density increases by a factor of roughly a thousand, meaning that the forces and stress encountered by Venus Express were much higher than during normal operations. The probe also experienced extreme

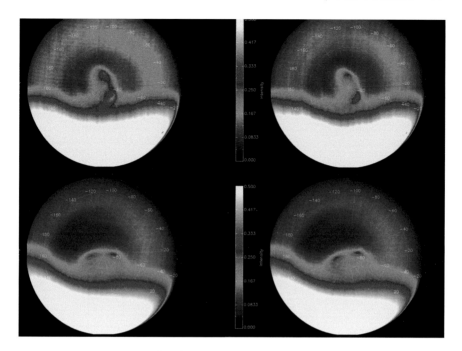

Fig. 5.7 Venus rotates slowly, yet it has permanent vortices in its atmosphere at both poles. These vortices form a constantly evolving structure on the surface of Venus. Long-term vortices are a frequent phenomenon in the atmospheres of fast rotating planets, like Jupiter and Saturn, for example. These *infrared* images show the changing shape of the double vortex over the south pole of Venus (Credit: ESA).

heating cycles, with temperatures rising by over 100 °C during several 100 s-long passages through the atmosphere.

Temperature and Seasons

Surface temperatures on Venus can rise to 482 °C, hot enough to melt lead, zinc and tin. The high temperature and pressure were responsible for the failure of many early space probes. Temperature and pressure in the atmosphere decrease with increasing altitude.

The dense atmosphere on Venus allows heat from the Sun to warm the surface but it also traps heat radiated from the surface of Venus. This results in a higher surface temperature than on Mercury (which is closer to the Sun). The trapping of heat by the

Fig. 5.8 Computer generated perspective image of Latona Corona and Dali Chasma, Venus. The image was created by superimposing Magellan radar data on topography vertically exaggerated by a factor of 10. The eastern part of the 1000 km diameter Latona Corona is on the *left*, and the view is from the *northeast* looking along the 3 km deep Dali Chasma (Credit: NASA).

atmosphere produces a greenhouse effect because the carbon dioxide acts like glass in a greenhouse. Earth has a greenhouse effect in its atmosphere, but Venus is an extreme case. The thick atmosphere of Venus also keeps the night side of Venus at nearly the same temperature as the side facing the Sun (unlike the night side of Mercury where the temperature drops dramatically). Temperatures at the poles of Venus are as hot as those at the equator.

As it orbits the Sun, Venus rotates very slowly on its axis, more slowly than any other planet. It takes 243 Earth days for just one spin, which means that a Venusian day is longer than a Venusian year.

Venus's rotation axis is tilted more than 177°, compared to Earth's 23.5° tilt. This means that Venus's axis is within 3° of being perpendicular to the plane of its orbit around the Sun.

Because of this, the planet has no seasons. Neither of the planet's hemispheres or poles point notably towards the Sun during any part of its orbit.

Magnetic Field

In 1967, Venera 4 found the magnetic field of Venus to be much weaker than that of Earths. This magnetic field is induced by an interaction between the ionosphere and the solar wind, rather than by an internal dynamo in the core like the one inside Earth. The lack of a strong magnetic field may be due to the slow rotation of Venus on its axis. One possibility is that Venus has no solid inner core, or that its core is not currently cooling, so that the entire liquid part of the core is roughly at the same temperature. Another possibility is that its core has already completely solidified.

Venus also has a weak magnetosphere that provides negligible protection to the atmosphere against incoming cosmic radiation. A weak magnetosphere means that the solar wind is interacting directly with the planet's atmosphere. Ions of hydrogen and oxygen are being created by the break up of neutral molecules by ultraviolet radiation. The solar wind then supplies energy that gives some of these ions sufficient velocity to escape Venus's gravitational field. This erosion process results in a steady loss of low-mass hydrogen, helium, and oxygen ions, whereas higher-mass molecules such as carbon dioxide are more likely to be retained. Atmospheric erosion by the solar wind probably led to the loss of most of Venus's water during the first billion years after it formed.

The ESA's Venus Express probe, which followed a near-polar orbit around Venus, also monitored the solar wind and Venusian magnetosphere. Data from the probe suggested Venus has a magnetotail but that it is much smaller than Earth's.

Further Information

http://solarsystem.nasa.gov (click on Venus)
www.space.com/venus/
www.nasm.si.edu (click on Venus)

6. Earth: The Planet of Life

Highlights

- The interaction of the Earth and the Moon slows the Earth's rotation by about 2 ms per century.
- In 2010 radar onboard India's Chandrayaan-1 space probe detected up to 600 million tonnes of water ice in craters at the Moon's north pole.
- In 2009, the LCROSS spacecraft was deliberately crashed into the lunar south pole. Water vapour and other molecules were detected in the debris plume.
- Recent space probes have mapped mass concentrations (mascoms) on the Moon's surface. In 2012, scientists determined how these mascons formed.
- In December 2013, China successfully carried out the world's first soft landing of a space probe on the Moon in nearly four decades.

Earth is the third planet from the Sun, orbiting at a average distance of 150 million km from the Sun. This distance is also known as one astronomical unit (1 AU).

Earth is the fifth largest planet with a diameter of 12,756 km. This makes Earth slightly larger than Venus. Earth orbits the Sun between Venus and Mars and is 1.4 times the distance from the Sun as Venus.

Earth is thought to have formed at the same time as the other planets in the Solar System about 4.5 billion years ago. Scientists known the age of the Earth because the oldest rocks ever discovered are 4.3 billion years old (determined from radioactive dating). Earth is the largest of the four rocky or terrestrial planets (Mercury, Venus, Earth and Mars). Like Mercury and Venus, Earth formed from a hot and molten state before it cooled to become a solid

J. Wilkinson, *The Solar System in Close-Up*, Astronomers' Universe,
DOI 10.1007/978-3-319-27629-8_6,
© Springer International Publishing Switzerland 2016

Fig. 6.1 The Earth as seen from a satellite about 36,000 km out in space. Water covers about 70 % of the Earth. Land covers only about 30 % (Credit: NASA).

planet. Even though it has cooled since formation, Earth is currently the most geologically active planet, and its interior is still quite hot. The main feature that separates Earth from the other planets in the solar system is that it is the only planet to contain water in the liquid state (Fig. 6.1 and Table 6.1).

Table 6.1 Details of Earth

Distance from Sun	149,600,000 km (1.0 AU)
Diameter	12,756 km
Mass	5.97×10^{24} kg (1.0 Earth's mass)
Density	5.52 g/cm^3 or 5520 kg/m^3
Orbital eccentricity	0.017
Period of revolution	365.3 Earth days (1.00 year)
Rotation period	1.0 Earth days (24 h)
Orbital velocity	107,244 km/h
Tilt of axis	23.5°
Day temperature	15 °C
Night temperature	10 °C
Number of Moons	1
Atmosphere	Nitrogen, oxygen
Strength of gravity	9.8 N/kg at surface

Early Views About Earth

Earth is the only planet whose English name has not been derived from Greek or Roman mythology. The name comes from Old English and Germanic. There are other names for Earth in other languages.

In Roman mythology, the goddess of the Earth was Tellus—the fertile soil. In Greek, Gaia means terra mater—Mother Earth.

Ancient understandings of Earth varied often according to religious views. For a long time people thought the Earth was flat because it seemed that way. It was not until the time of Copernicus in the sixteenth century that it was understood that the Earth was just another planet.

Views of the Earth from space probes have confirmed that it is indeed a spherical body.

Probing Earth

Human exploration of the solar system began with the Earth. The first spacecraft were small-unmanned craft launched into the Earth's atmosphere. Improvements in space technology eventually led to manned craft orbiting Earth in a period of time known as the 'space race'.

Early in the space race the USSR was active with its Sputnik, Vostok, Voskhod and Soyuz spacecraft. Early USA spacecraft placed in Earth orbit included Explorer and those of the Mercury and Gemini programs. These early missions were the pioneers of future exploration of the Moon and other planets. For example, the 10 manned Gemini missions between 1964 and 1966 involved rendezvous between spacecraft in orbit, space walks, and even dockings with unmanned target vehicles. The American Apollo program contained many spacecraft missions that eventually resulted in humans visiting the Moon (Earth's only satellite).

Today there are thousands of satellites in orbit around Earth, each belonging to particular countries. These are being used for many purposes, including communications, defence, science, GPS navigation, map making, environmental and weather monitoring. Orbiting observatories, such as the International Space Station are the largest and most complicated of all scientific satellites. Such satellites contain many types of scientific instruments that provide valuable research data.

Most Earth observation satellites orbit above 500 km in order to avoid the significant air-drag that occurs at low altitudes. The Earth observation satellites ERS-1, ERS-2 and Envisat of European Space Agency as well as the MetOp spacecraft of the European Organisation for the Exploitation of Meteorological Satellites are all operated at altitudes of about 800 km. Many are operated in a Sun synchronous polar orbit so as to get better global coverage. Such an orbit will have a period of roughly 100 min.

Spacecraft carrying instruments (e.g. meteorological satellites) for which an altitude of 36,000 km is suitable sometimes use a geostationary orbit. Such an orbit allows uninterrupted coverage of more than 1/3 of the Earth. Three geostationary spacecraft at longitudes separated with 120° can cover the whole Earth except the extreme polar regions.

The Landsat program of NASA offers the longest continuous global record of the Earth's surface (over 40 years); it continues to deliver visually stunning and scientifically valuable images of our planet. Currently in operation are Landsat 7 and 8.

The National Oceanic and Atmospheric Administration of the USA (NOAA) has a number of environmental satellites that monitor the Earth from space. They are also used to analyse the coastal waters, relay life-saving emergency beacons, and track tropical storms and hurricanes. NOAA satellites also monitor conditions in space and solar flares from the Sun help us understand how conditions in space affect the Earth. The NOAA system uses both polar and geostationary satellites.

The Geostationary Operational Environmental Satellite system (GOES), operated by the United States National Environmental Satellite, Data, and Information Service, supports weather forecasting, severe storm tracking, and meteorology research. The National Weather Service (NWS) uses the GOES system for its United States weather monitoring and forecasting operations, and scientific researchers use the data to better understand land, atmosphere, ocean, and climate interactions. The GOES system uses geosynchronous satellites. Currently in operation are GOES 13, 14 and 15.

NASA's Earth Observing System (EOS) is a coordinated series of polar-orbiting and low inclination satellites for long-term global observations of the land surface, biosphere, solid Earth, atmosphere, and oceans. As a major component of the Earth Science Division of NASA's Science Mission Directorate, EOS enables an improved understanding of the Earth as an integrated system. More information can be found on their website. On 2 July 2014, NASA launched the Orbiting Carbon Observatory-2 (OCO-2) mission. OCO-2 is the first NASA satellite dedicated to monitoring carbon dioxide, and it will do so with greater precision and detail than current instruments.

The UCS Satellite Database is a listing of the more than 1000 operational satellites currently in orbit around Earth. It includes basic information about the satellites and their orbits, but does not contain the detailed information necessary to locate individual satellites.

Position and Orbit

Earth orbits the Sun in a slightly elliptical orbit. Its mean distance from the Sun is just over 149 million km. At closest approach (perihelion) it is 147 million km from the Sun and its furthest distance (aphelion) is 152 million km.

The Earth like the other planets has three motions. It spins like a top on its axis, it travels around the Sun, and it moves through the Milky Way galaxy with the rest of the Solar System. The Earth's axis is an imaginary line that connects the North and South poles. The spinning motion causes day and night on Earth, since only one side faces the Sun at any one time.

The Earth takes just over 365 days, 6 h and 9 min to orbit the Sun once; this length of time is called a sidereal year. Its axis is inclined at 23.5° off the perpendicular to the plane of its orbit. A single rotation on its axis takes 23 h 56 min and 4 s; this length of time is called a sidereal day. Earth rotates from west to east on its axis and this makes the Sun appear to move across the sky from east to west each day. The tilt of the Earth's axis and its position around the Sun cause the seasons.

The interaction of the Earth and the Moon slows the Earth's rotation by about 2 ms per century. Research indicates that about 900 million years ago there were 481 eighteen-hour days in a year.

Density and Composition

The Earth has a slightly larger mass, diameter and average density than Venus. Because of this, the strength of gravity on Earth is slightly more than that of Venus. A 75 kg person on Earth would weigh 735 N, but on Venus they would weigh only 607 N. Earth is also the densest planet in the Solar System, due to the fact that it has a large nickel-iron core.

Scientists know a lot about the interior of the Earth from their study of earthquake (seismic) waves and volcanoes. Earthquakes release enormous amounts of energy that travel through the Earth and along its surface as waves.

The interior of the Earth is divided into three main layers, which have distinct chemical and seismic properties: the core, mantle and crust. The core has an inner and outer layer. The inner core makes up only 1.7 % of the Earth's total mass and is solid iron and nickel. It is extremely hot (about 5000 °C) but remains a solid because of the pressure from surrounding layers. At the centre the pressure is about four million times greater than at the surface. Surround the inner core is a liquid outer core containing about 30 % of the Earth's mass. The outer core has a temperature of about 4100 °C. Convection currents in the liquid outer core are thought to produce electrical currents that generate the Earth's magnetic field.

The outer core is surrounded by the mantle, which contains 67 % of the Earth's mass. The mantle is 2820 km thick and contains iron/magnesium silicate minerals and oxygen that are kept in the solid state by high pressures. Nearer the surface of the mantle there are some molten regions that are liquid enough to flow. Sometimes this molten material rises to the surface and forms volcanoes and lava flows.

Surrounding the mantle is a thin outer layer or crust about 40–70 km thick that accounts for about 0.4 % of the Earth's mass. It is made of granitic and basaltic rocks, which contain silicon dioxide (quartz) and other silicates like feldspar. The surface of the crust is covered by oceans (70 %) and continental land (30 %).

The crust varies in thickness. It is thicker under the continents and thinner under the oceans (Table 6.2 and Fig. 6.2).

Unlike the other terrestrial planets, Earth's crust is divided into several separate solid plates, which move around on top of the hot mantle material. The continents are attached to the plates. Some plates are moving towards each other, and when they collide, material builds up to form fold mountains (such as the Himalayas). When plates are moving apart, molten material from the mantle often gets drawn up creating new crust; this occurs along the mid-Atlantic ridge between South America and Africa. Other plates are sliding past each other in a slow but jerky motion; such as the San Andreas Fault in California. Sometimes an oceanic plate slides underneath a continental plate, causing a deep trench (subduction zone) and mountain ranges; the Andes Mountains in

Table 6.2 Chemical composition of the Earth

Element	Proportion (%)
Iron	34.6
Oxygen	29.5
Silicon	15.2
Magnesium	12.7
Nickel	2.4
Sulfur	1.9
Titanium	0.05

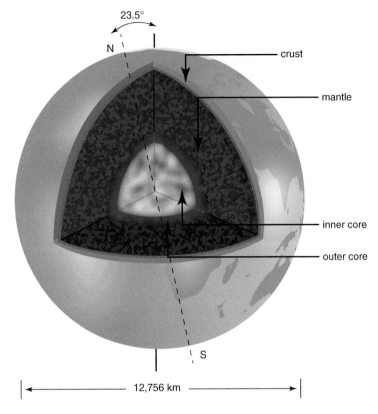

Fig. 6.2 The interior structure of Earth.

South America were formed this way. Where plates interact, earth-quake and volcanic activity occurs along the boundaries. Some volcanoes and earthquakes do not occur along plate boundaries, but in places are called 'hot spots'—this accounts for the occa-sional earthquake in Australia, which rides in the middle of a plate.

The slow movement of plates (a few centimetres a year) is driven by convection currents and tectonic forces in the mantle (see Fig. 6.3).

The Surface

The crust of the Earth is a thin layer containing rock material.

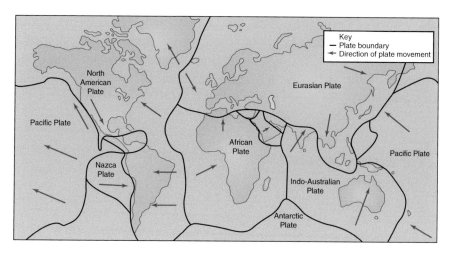

Fig. 6.3 Map showing the major plate boundaries of Earth.

The surface of the crust consists of mountain ranges, valleys, flat plains, deserts, and vegetated areas of varying density. From space, it can be seen that oceans of water cover much of the surface. The landscape has been shaped over millions of years by tectonic forces and volcanic action and erosion by the wind, water and glaciers.

The four terrestrial planets—Mercury, Venus, Earth and Mars—contain evidence of volcanic action having occurred in their past. In the case of Earth, some volcanoes are still active and these often erupt, releasing molten rock material into the air or surrounding areas. There are about 500 volcanoes on Earth, most are found along plate boundaries.

A key feature of the Earth is the presence of liquid water on its surface. This water is found in the oceans, lakes, and rivers. Water also occurs in the polar ice-caps, and as vapour in the air. Water allows living things to survive on Earth.

The highest land feature on Earth is Mount Everest at 8848 m above sea level. The lowest land feature is the shore of the Dead Sea at 399 m below sea level. The deepest part of the ocean is an area in the Mariana Trench in the Pacific Ocean southwest of Guam, 11,030 m below the surface. Average ocean depth is 3795 m.

The largest impact structure discovered on Earth is the Chicxulub Crater. It is hidden under sediments on the coast of the Yucatan Peninsula. The crater is a circular structure about 180 km wide and was discovered when instruments detected variations in the Earth's gravitational and magnetic field. The meteorite that made the structure has been estimated to be about 10 km across.

The Atmosphere

The Earth's atmosphere contains 7 % nitrogen, 21 % oxygen, with traces of argon, carbon dioxide and water.

In recent years the amount of carbon dioxide in the atmosphere has been increasing, due mainly to the burning of fossil fuels. At the same time people are cutting down forest trees that use carbon dioxide. The increase in carbon dioxide levels has resulted in higher average temperatures via the greenhouse effect. Carbon dioxide gas allows sunlight to reach the surface but prevents heat from escaping, in this way the atmosphere warms up. However, without the greenhouse effect surface temperatures would be much colder than they currently are.

Oxygen began to accumulate in the atmosphere when primitive life forms (first bacteria and plants) began to photosynthesis. Increases in oxygen levels allowed more forms of life to evolve.

Only a small amount of heat given off by the Sun enters the atmosphere; most is lost in space. About 34 % of the sunlight that enters the atmosphere is reflected back into space by clouds. Only 19 % of the sunlight that enters the atmosphere heats it directly. The remaining 47 % heats the ground and seas. Re-radiated heat from the ground and seas causes most of the warming of the atmosphere.

Temperature and Seasons

The Earth is heated by the Sun and has an average surface temperature of 15 °C. The highest recorded temperature on the surface is 58 °C at Al Aziziyah in Libya; the lowest temperature recorded is −89.6 °C at Vostok Station in Antarctica.

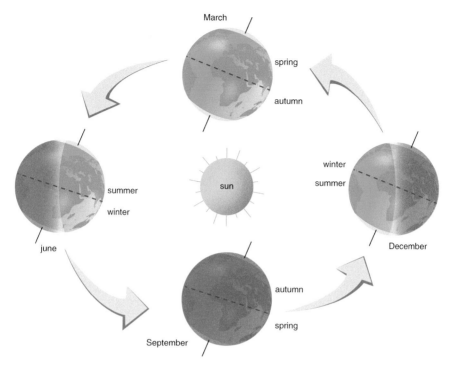

Fig. 6.4 The seasons on Earth.

Varying temperatures and pressures in the atmosphere create strong winds.

The Earth has seasons because of its slightly elliptical orbit around the Sun and the fact that its axis is tilted at 23.5 °C off the perpendicular to the plane of its orbit (see Fig. 6.4).

The four climatic seasons are called summer (hot conditions), autumn (mild conditions), winter (cold conditions) and spring (mild conditions). The seasons in the northern hemisphere are opposite to those in the southern hemisphere. Today the Earth has a fairly stable climate with a narrow range of temperatures. In the past, however, there may have been rapid and dramatic climate change. Such changes may have resulted from changes in the position of the Earth in its orbit around the Sun, increased volcanic activity, changed atmospheric composition or large meteorite impacts.

Magnetic Field

The Earth has a modest magnetic field produced by electric currents in its iron and nickel core. The magnetic field forms a set of complete loops similar in shape to those around a bar magnet. The magnetic poles of the Earth are close to the rotational pole (about 11° apart); other planets have much larger angles between their magnetic and rotational poles.

The Earth's magnetic field (magnetosphere) normally protects it from bombardment by energetic particles from space. Most of these particles originate from the Sun and are carried towards Earth by the solar wind. Near the Earth the solar wind has a speed of about 400 km/s. Charged particles in the solar wind get trapped in the magnetic field and circulate in a pair of doughnut-shaped rings called the **Van Allen radiation belts**. These belts were discovered in 1958 during the flight of America's first successful Earth-orbiting satellite named after physicist James Van Allen, who insisted the satellite carry a Geiger counter to detect charged particles (see Fig. 1.8).

During times of solar flares and increased sunspot activity, the Van Allen Radiation belts become overloaded with charged particles. Some of the particles travel down into the Earth's upper atmosphere where they interact with gases like nitrogen and oxygen to cause them to fluoresce (give off light). The result is a shimmering display called the northern lights (aurora borealis) or southern lights (aurora australis) depending on the hemisphere involved. Such events sometimes interfere with radio transmissions, communication satellites, and electrical power transmission.

The Moon

The Earth has only one natural satellite—the Moon. The Moon orbits Earth in an elliptical orbit at about 36,800 km/h. Its from Earth varies from 356,000 to 407,000 km. The Moon is about a quarter the size of Earth and has just more than 1 % of the mass of

Fig. 6.5 The Moon as imaged by the author through a 10 inch SCT tele-scope in April 2015 (Credit: J. Wilkinson).

Earth. The density of the Moon is only 3.34 g/cm^3 compared to Earth's 5.52 g/cm^3 (Fig. 6.5).

The Moon is the second brightest object in the sky after the Sun. Unlike the Sun, which emits its own light and heat, the Moon only reflects sunlight. The Moon is also one of the most widely studied objects in the solar system. It has been studied with the naked eye, telescopes, and spacecraft. To date, the Moon is the only body in the Solar System to have been visited by humans; this occurred for the first time in 1969.

The Moon was probably formed at the same time as other planets in the solar system and gravitationally captured by the Earth. Others think it is a fragment torn out of Earth's mantle. Yet another theory is that the Earth–Moon pair could be a double planet.

Early Views About the Moon

The presence of the Moon in our sky has captured human interest throughout history. The Moon is so large and close to Earth that some of its surface features are readily visible to the naked eye. For centuries people have thought that some of the features on the Moon looked like a human face looking down on them, and often talked about the 'man in the Moon'. People also thought that the Moon had a fairly smooth surface, until Galileo's telescope showed the surface was covered with many craters and mountains as well as plains (Table 6.3).

Probing the Moon

More spacecraft have been sent to the Moon than any other body in the solar system, simply because it is so close to Earth. Early probes to pass by the Moon included the USSR's Luna probes and the USA's Ranger and Surveyor probes.

The most significant missions to the Moon were those of the USA's **Apollo program**. The first manned lunar fly-around involved 10 orbits of the Moon by Apollo 8, between 21 and 27 December 1968. Apollo 11 was the first manned lunar landing, on 20 July 1969. The landing site was Mare Tranquillitatis. Neil Armstrong was the first astronaut to walk on the Moon followed

Table 6.3 Details of the Moon

Distance from Earth	384,400 km
Diameter	3476 km
Mass	7.35×10^{22} kg (0.012 Earth masses)
Density	3.34 g/cm^3 or 3340 kg/m^3
Orbital eccentricity	0.055
Period of revolution	27.3 Earth days (relative to stars)
	29.5 Earth days (seen from Earth)
Rotation period	27.3 Earth days
Orbital velocity	36,800 km/h
Tilt of axis	6.7°
Day temperature	130 °C
Night temperature	−184 °C
Atmosphere	None
Strength of gravity	1.7 N/kg at surface

Fig. 6.6 View of the Moon's surface, the Apollo 11 lander, and distant Earth (Credit: NASA).

by Edwin Aldrin. The third crew member, Michael Collins, stayed in the Command module orbiting the Moon (Figs. 6.6 and 6.7).

Apollo missions 12 and 14 through to 17 landed manned craft on the Moon's surface. Of these missions, Apollo 15 was first to make use of a lunar rover vehicle. The rover allowed astronauts to travel several kilometres from the landing site. The last Apollo mission (Apollo 17) occurred in December 1972.

Exploration of the Moon continued with the Galileo and Clementine spacecraft. In October 1989, Galileo began a 6-year mission to Jupiter, but on its way passed the Moon twice. Data was returned to Earth on the composition of the lunar surface. Clementine went into orbit around the Moon in February 1994. Using

Fig. 6.7 Apollo 15 astronaut, lander and rover on the Moon's surface (Credit NASA).

laser-ranging techniques and high-resolution cameras, Clementine mapped the Moon's topography in greater detail than had been done previously. In 1999, Prospector also mapped the lunar surface and detected ice buried beneath the ground in deep polar craters. More recently Japan, India and China have sent space probes to the Moon.

These recent missions have been successful at 3D mapping of the Moon, spectral analysis of the surface and interior, measurement of the gravitational field, surveying lunar resources and identifying future landing sites (see Table 6.4).

In December 2013, China successfully carried out the world's first soft landing of a space probe on the Moon in nearly four decades. The touchdown of the unmanned Chang'e 3 lander was the latest mission in the country's ambitious space programme, which is intended to put a Chinese astronaut on the Moon early next decade. The lander carried a six-wheeled rover called Yutu (Jade Rabbit), to its landing place on a flat plain known as the Sinus

Table 6.4 Recent space probes to the Moon

Spacecraft	Country of origin	Date	Mission focus
SMART-1	Europe	2004	Lunar geology
Selene (Kaguya)	India	2007	Minerals, geographic
Chang'e 1	China	2007	Mapping, geology
Chandrayaan 1	India	2008	Spectral analysis
Moon Impact Probe	India	2008	Close range imaging
Lunar Recon. Orbiter	USA	2009	Resources, mapping
Chang'e 2	China	2010	Imaging, analysis
ARTEMIS P1	USA	2011	Monitor solar wind
ARTEMIS P2	USA	2011	Monitor solar wind
GRAIL A	USA	2011	Measure gravity
GRAIL B	USA	2012	Measure gravity
LADEE	USA	2013	Lunar exosphere
Chang'e 3	China	2013	Soft lander, rover

Iridum, or Bay of Rainbows, after hovering over the surface for several minutes before selecting the best available landing spot. This region was selected for the craft's landing because lunar probes had not previously studied it. The spaceship's rover was remotely controlled by Chinese control centres with support from a network of tracking and transmission stations around the world operated by the European Space Agency. After it touched down on the Moon, Chang'e 3's solar panels, which are used to generate power from sunlight, unfolded and the spacecraft began transmitting pictures back to Earth.

India expects to launch another lunar mission by 2016, which would place a motorised rover on the Moon. China plans to conduct a sample return mission with its Chang'e 5 spacecraft in 2017.

The Japanese Aerospace Exploration Agency (JAXA) plans a manned lunar landing around 2020 that would lead to a manned lunar base by 2030.

Position and Orbit

The Moon spins like a top on its axis as it travels around the Earth, and, at the same time the Earth is orbiting the Sun. The Moon's orbit is slightly elliptical and its mean distance from the Earth is

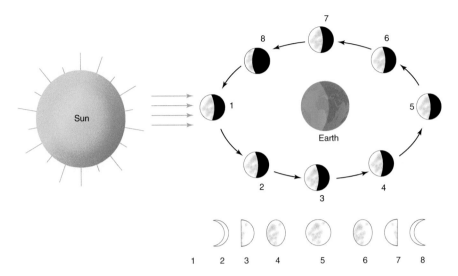

Fig. 6.8 Phases of the Moon.

384,400 km. At closest approach (perigee) it is 356,000 km from Earth and its furthest distance (apogee) is 407,000 km.

The Moon takes 27.3 days to go once around the Earth and it also takes this time to rotate once on its own axis. Because of this, we always see the same side of the Moon from Earth. As the Moon orbits the Earth, the angle between the Earth, Moon and Sun changes, and we see this as the cycle of Moon's phases (see Fig. 6.8). The time between successive new Moons is actually 29.5 days, slightly different from the Moon's orbital period (measured against the stars) because the Earth moves a significant distance in its orbit around the Sun in that time.

The Moon appears to wobble on its axis due to its elliptical orbit. As a result we can see a few degrees of the far side of the Moon surface from time to time. Most of the far side was completely unknown until the Soviet spacecraft Luna 3 photographed it in 1959. The far side of the Moon gets sunlight half the time. Whenever we see less than a full Moon, some sunlight is falling on the far side. Throughout each cycle of lunar phases all parts of the Moon get equal amounts of sunlight.

Gravity keeps the Moon in orbit around the Earth and produce the tides. Tidal forces deform the oceans, causing them to rise at some places and to settle elsewhere. There are two areas of high tide on the Earth at any given time, one on the side closest to the Moon, the other on the opposite side of the Earth. Low tides occur where the Moon is on the horizon.

The Sun also distorts the shape of the oceans, but only half as much as the Moon, because the Sun is nearly 400 times farther away.

Instruments placed on the Moon by the Apollo 12 astronauts have enabled us to measure precisely the distance to the Moon. From such measurements over time, astronomers have found that the Moon is spiralling away from the Earth as a rate of about 4 cm per year. The cause of this motion is thought to be due to tidal interactions between the Moon and Earth. Tidal forces are also causing the Earth's rotation to slow by about 1.5 ms per century.

Density and Composition

The Moon has a smaller mass, diameter and average density than Earth. Because of this, the strength of gravity on the Moon's surface is one sixth that of Earth. A 75 kg person on Earth weighs 735 N, but on the Moon the same person would only weigh 122 N. The Moon's escape velocity is only 2.4 km/s.

Scientists have found out about what the interior of the Moon is like from seismic (moonquake) evidence made at the Apollo landing sites. The Moon has a crust, mantle and core. Although these regions are similar to those of the Earth, the proportions are quite different. The Moon's crust averages 68 km thick and varies from a few kilometres under Mare Crisium on the visible side, to 107 km near the crater Korolev on the far side. The GRAIL mission showed the lunar crust is thicker on the far side (60 km) than on the nearside (20–30 km). This means the Moon is slightly egg-shaped, with the small end pointing toward Earth. As a result, the Moon's centre of mass is 2 km closer to Earth than its geometric centre. The difference in thickness also helps explain why most of the mare basalt lavas are confined to the near side of the Moon. On the near side of the Moon the lavas would have reached the surface more easily.

Seismometers placed on the Moon by the Apollo astronauts found moonquakes occur on the Moon on a fairly regular cycle of about 2 weeks. They apparently result from the tidal stresses induced in the Moon as it rotates about Earth. Most moonquakes measure less than three on the Richter scale.

Below the crust is the Moon's mantle. The Moon's mantle probably makes up most of its interior. Unlike the Earth's mantle however, the Moon's mantle is only partially molten.

The mantle is solid down to a depth of about 800–1000 km. The composition of the upper mantle may be deduced from the composition of the mare lavas, which came from these regions. Below about 1000 km the mantle becomes partially molten. Evidence for this came mainly from seismic data collected when a large meteorite weighing about one tonne hit the far side of the Moon in July 1972. At the centre of the Moon there may be a small iron-rich core perhaps only 800 km in diameter, but its existence is uncertain.

Analysis of Moon rocks shows no evidence for formation in a different part of the solar system from Earth. There is some evidence to support the theory that the Moon may have been once part of the Earth. The bulk density of the Moon is close to the silicate mantle of the Earth; however the bulk composition of the Moon is different to that of Earth's mantle. The Moon as a whole contains a higher proportion of iron and a lower proportion of magnesium than the Earth's mantle. The Moon also has a lower proportion of lead but a higher proportion of calcium, aluminium and uranium than Earth.

Information from the Moon rocks support an 'impact theory' for the formation of the Moon. In this theory a large object, about the size of Mars, hit Earth in its first 100 million years of life. This collision literally ejected a lot of rock material from the Earth's surface into orbit forming a 'debris ring'. This ring gradually condensed into the Moon. Such an impact could also have tipped the Earth off its axis and so created the seasons.

The Surface

There are two main types of terrain on the Moon and these can be identified by the naked eye—the heavily cratered and very old highlands or terrae, and the relatively smooth and younger plains or maria (singular mare). Most of the surface is covered with regolith, a mixture of fine dust and rocky debris produced by meteor impacts. This layer ranges in thickness from 1 to 20 m. Unlike Earth's soil, which has decayed biological matter in it, the Moon's regolith does not have any biological matter.

The maria make up about 16 % of the Moon's surface and are sometimes called 'seas' even though they contain no water (mare means 'sea' in Latin). Maria are huge impact basins that have been covered by molten lava. Most maria exist on the side of the Moon facing Earth. The more prominent mare are Mare Tranquillitatis (Sea of Tranquillity), Mare Nubium (Sea of Clouds), Mare Nectaris (Sea of Nectar), and Mare Serenitatis (Sea of Serenity). The largest mare, Mare Imbrium (Sea of Showers) is circular and measures 1100 km in diameter. Like most maria, it is 2–5 km below the average lunar elevation.

Rocks brought back to Earth from the maria are solidified lava (mainly basalt), which suggest the Moon's surface was once molten. These rocks have a composition similar to those found in volcanic rocks on Hawaii or Iceland—they contain heavy elements like iron, manganese and titanium. The molten lava has come from inside the Moon and has risen to the surface through large impact fractures in the crust.

There is also some tectonic activity in the maria caused by the weight of basalts pushing on the crust. At the edges of the maria, the basalts are stretched, causing fracturing and faulting. In the interior of maria, the basalts are compressed, resulting in folding that produces wrinkle ridges. Most mare also contain small craters and occasional cracks (lava tunnels or channels) called rilles. In the highlands, tectonic activity has produced small scarps.

Lunar probes have shown that the far side of the Moon contains one prominent mare, Mare Moscoviense, and is heavily cratered. The cratered area on the far side is 4–5 km above the average lunar elevation.

The Moon's surface is covered with meteorite impact craters that vary in size from tiny pits to huge craters hundreds of kilometres in diameter. Virtually all the craters are round and the result of meteorite impact. Some of the craters have rays or streaks extending outwards from their centre, while others have raised peaks at their centre. These peaks occur because the impact compresses the crater floor so much that afterwards the crater rebounds and pushes the peak upwards. As the peak goes up, the crater walls collapse and form terraces. One of the most striking craters with a central peak is Copernicus (see Fig. 6.10). Copernicus crater is about 92 km across and 800 million years old. Rays are often formed when material is ejected and scattered across the surface during large impacts. The most striking crater with rays is Tycho formed about 109 million years ago (see top crater in Fig. 6.5).

Most of the craters on the near side are named after famous figures in the history of science, such as Tycho, Copernicus and Kepler. Other craters bear the names of philosophers such as Plato and Archimedes. Features on the far side have modern names such as Apollo, Gagarin and Korolev, with a distinctly Russian bias, since the first images of the far side were obtained by Luna 3. The largest impact basin (crater) is the Aitken basin, 2500 km wide and 12 km deep, at the south pole on the far side. The Imbrium basin, about 1800 km wide and the Crisium basin, about 1100 km wide, are both found on the near side.

In contrast to the dark maria, the light-coloured highlands are elevated regions that make up about 84 % of the lunar surface. Some of the highlands are mountains or ridges that form the rims of large basins that were formed from material uplifted after impacts. One of the highest mountains on the Moon, the Apennines, forms part of the Imbrium basin (see Fig. 6.9).

Highland rocks are light anorthosites (feldspars) rich in calcium and aluminium. Many highland rocks brought back to Earth are impact breccias, which are composites of different rocks fused together as a result of meteorite impacts.

Radioactive dating of Moon rocks have shown the mare rocks to be between 3.1 and 3.8 million years old, while the highland rocks are between 4.0 and 4.3 billion years old.

In September 2009, India's ISRO Chandrayaan-1 detected water-ice on the Moon and hydroxyl absorption lines in reflected sunlight. In March 2010, it was reported that the Mini-RF on board the Chandrayaan-1 probe had discovered more than 40 permanently darkened craters near the Moon's north pole that are

Fig. 6.9 The Apennines are the largest mountain range on the Moon. Notice the wrinkle ridges and the large crater Archimedes with the lava flooded *bottom*. (Credit: J. Wilkinson)

thought to contain an estimated 600 million metric tonnes of water-ice.

In November 2009, NASA reported that its LCROSS space probe had detected a significant amount of hydroxyl in the material thrown up from a south polar crater by an impactor; this may be attributed to water-bearing materials such as pure crystalline water-ice. The suite of LCROSS and LRO instruments determined as much as 20 % of the material kicked up by the LCROSS impact was volatiles, including methane, ammonia, hydrogen gas, carbon dioxide and carbon monoxide. The instruments also discovered relatively large amounts of light metals such as sodium, mercury and possibly even silver. The science team at NASA found the water-ice on the Moon is not uniformly distributed within the shadowed cold traps, but rather is in pockets, some of which may lie outside the shadowed regions.

Fig. 6.10 Copernicus crater (*large crater on left*) and Kepler crater (*shaded one on right*) (Credit: J. Wilkinson).

Scientists believe water may have been delivered to the Moon over geological timescales by the regular bombardment of water-bearing comets, asteroids and meteoroids or continuously produced in situ by the hydrogen ions (protons) of the solar wind impacting oxygen-bearing minerals. The search for the presence of lunar water has attracted considerable attention and motivated several recent lunar missions, largely because of water's usefulness in rendering long-term lunar habitation feasible.

Large concentrations of mass lurk under the lunar surface. These concentrations change the gravity field and can either pull a spacecraft in or push it off course. The GRAIL missions have now mapped where theses areas are, and scientists have a much better understanding of how they developed. In 2012, the GRAIL team confirmed the theory that the concentrations of mass (called Mascons) were caused by massive asteroid impacts billions of years ago. The researchers determined that ancient asteroid impacts excavated large craters on the Moon, causing surrounding lunar materials and rocks from the Moon's mantle to melt and collapse inward. This melting caused the material to become denser and more concentrated.

In 2014, scientists studying GRAIL data reported that they have found evidence that the craggy outline of the Oceanus Procellarum region of the Moon is actually the result of the formation of ancient rift valleys. The rifts are buried beneath dark volcanic plains and are unlike anything found anywhere else on the Moon. Another theory arising from the data analysis suggests this region formed as a result of churning deep in the interior of the Moon that led to a high concentration of heat-producing radioactive elements in the crust and mantle of this region.

The Atmosphere of the Moon

The Moon has no real atmosphere and no liquid water so life could not exist for long. Atmospheres are held in place by gravity, and the Moon has so little gravitational pull that it is unable to hold any of the gases such as those that make up the Earth's atmosphere.

The LADEE spacecraft orbited close to the surface of the Moon during the first few months of 2014 and found a veil of micron-size dust particles continuously encases the Moon. The rain of meteorite matter that hits the surface kicks up these particles. The spacecraft picked up helium, neon, and argon in the Moon's tenuous, transient atmosphere, and it detected atoms of magnesium, aluminium, titanium, and oxygen—the remnants of rocky mineral blasted upwards from the lunar surface. LADEE smacked into the farside at 5700 km/h on 17 April 2014.

Temperature

There is a range of temperatures on the surface of the Moon, because of its lack of atmosphere. At night temperatures can fall to −184 °C, while on parts of the Moon facing the Sun temperatures can reach 130 °C. At the poles, temperatures are constantly low, about −96 °C. Some polar regions are in permanent shadow.

Because the Moon rotates once on its axis every 27.3 days, night and day at any point on the Moon last about 14 Earth days. On the side of the Moon that always faces Earth, 'phases of the Earth' would be observed. Part of the long period of night would be 'Earth lit', just like we have 'Moon lit' nights on Earth.

Magnetic Field

The Moon has no global magnetic field, but some rocks brought back by the Apollo astronauts exhibit permanent magnetisms. This suggests that there may have been a global magnetic field early in the Moon's history.

With no atmosphere and no magnetic field, the Moon's surface is directly exposed to the solar wind. Since the Moon's early days many charged particles from the solar wind would have become embedded in the Moon's regolith (surface material). Samples of regolith returned by the Apollo astronauts confirmed the presence of these charged particles.

Further Information

http://nssdc.gsfc.nasa.gov (click on the Moon)
www.moon.nasa.gov
For specific information about the Moon and its features see the book: "The Moon in Close Up", written by John Wilkinson and published by Springer (2010).

7. Mars: The Red Planet

Highlights

- The two rovers Spirit and Opportunity have found that water once existed on the Martian surface in its past.
- The Martian surface contains many large basins formed when large meteors or asteroids have hit the surface.
- The soil on Mars is mostly iron-rich clay that contains iron, silicon and sulfur; and it is slightly magnetic.
- The Curiosity probe that landed in Gale crater has found evidence that rivers and streams once flowed over the crater floor.
- Small amounts of water vapour in Mar's atmosphere can form fogs and clouds.
- The two moons of Mars are probably captured asteroids.

Mars is regarded as one of Earth's neighbours in space. Many people have considered Mars to be the most likely planet, apart from Earth, to contain life. Mars is the fourth planet from the Sun, orbiting at an average distance of 228 million km. This distance is about one and a half times the distance Earth is from the Sun. Radio signals take between 2.5 min and 20 min to travel one way between Earth and Mars, depending on where Mars is in its orbit in relation to Earth. At times Mars is the third brightest planet we see in our night sky, after Venus and Jupiter. It has a diameter of 6794 km, about half Earth's diameter, making it the seventh largest planet. Mars is often referred to as 'the red planet' because it appears red from Earth. This colour is due to the large amounts of red dust that cover its surface. The planet is thought to have formed about 4.5 billion years ago, at the same time as the other planets in the solar system. Because it is relatively close to the Sun, Mars must have been hot and in a molten state before it cooled to become a solid planet (see Fig. 7.1).

J. Wilkinson, *The Solar System in Close-Up*, Astronomers' Universe,
DOI 10.1007/978-3-319-27629-8_7,
© Springer International Publishing Switzerland 2016

Fig. 7.1 A view of Mars as seen by Mars Global Surveyor. Notice the three volcanoes on the *right* hand side and the "Valles Marineris" canyon across the *middle* of the image (Credit: NASA).

Early Views About Mars

Mars is named after the ancient Roman god of war. Both the ancient Greeks and Romans associated Mars with war because its colour resembles that of blood. The Greeks called the planet Ares. The two moons of Mars, Phobos (fear) and Deimos (panic) are named for the sons of the Greek god of war. The month of March, derives its name from Mars.

Mars has been known since prehistoric times and many ancient astronomers have studied its motion in the night sky. The Danish astronomer Tycho Brahe (1546–1601) made decades of observations of Mar's motion. Tycho's work was continued by Johannes Kepler (1571–1630) who used the observations to develop the first two of his three laws of planetary motion. These included the conclusion that planets orbit in an elliptical path, with the Sun at one focus.

Another early astronomer to study Mars was the Italian-French Giovanni Domenico Cassini who, in 1666, made the first reasonably accurate measurements of Mars's axial rotation period, which he found to be 37.5 min longer than that of the Earth. Cassini was also the first to observe the Martian polar ice caps. The first observations of the surface markings were made in 1659 by Christiaan Huygens who drew the dark triangular feature we now know as the large plateau Sytris Major. Huygens also estimated the length of the Martian day to be about 24 h.

In 1777, William Herschel measured the tilt of the axis of Mars and deduced it must have seasons like Earth because it underwent regular changes in its polar ice caps.

Wilhelm Beer and Johann von Madler made the first detailed maps of the Martian surface features during the 1830s.

In 1877 the Italian astronomer Giovanni Schiaparelli reported seeing 'canali' (channels) on the surface of Mars. American Percival Lowell reported seeing similar features in the early 1900s. Lowell thought Mars was a desert world and that inhabitants of Mars used the features to carry water from the ice caps to equatorial regions. Exploration of Mars by space probes and high-resolution telescopes has since disproved the existence of such features (Table 7.1).

Table 7.1 Details of Mars

Distance from Sun	227,940,000 km (1.52 AU)
Diameter	6794 km
Mass	6.42×10^{23} kg (0.107 Earth's mass)
Density	3.95 g/cm^3 or 3950 kg/m^3
Orbital eccentricity	0.093
Period of revolution	687 Earth days
Rotation period	1.029 Earth days
Length of year	1.881 Earth years
Orbital velocity	86,868 km/h
Tilt of axis	25.2°
Average temperature	−60 °C
Number of Moons	2
Atmosphere	Carbon dioxide
Strength of gravity	3.6 N/kg at surface

Probing Mars

People on Earth have observed Mars through telescopes based on Earth and in space. Early space probes carried telescopes and cameras to observe Mars as they flew past it. Later probes went into orbit around Mars and collected much more data. More recently, probes have successfully landed on the Martian surface, but to date, no human has yet set foot on Mars.

The first probe to fly by Mars was Mars 1, launched by the USSR on 1 November 1962, failed to return data. Between 1965 and 1969 the USA's Mariner 4, 6 and 7 probes passed by Mars and took many photographs of the desert-like surface. The thin atmosphere was confirmed to be composed of carbon dioxide, and a weak magnetic field was detected.

The Mars 2 space probe reached Mars in 1971 and released a lander that crashed into the Martian surface when its rockets failed to slow it down. No data was returned but it was the first human-made object to be placed on Mars. The Mars 3 probe arrived at Mars on 2 December 1971 and a lander was successfully placed on the surface, however it returned video data for only 20 s.

The first US spacecraft to enter an orbit around Mars was Mariner 9 on 3 November 1971. At the time of its arrival a huge dust storm was in progress on Mars and many experiments had to be delayed until the storm had finished. The probe sent back the first high-resolution images of the moons, Phobos and Deimos. Mariner 9 took over 7000 images of Mars and showed ancient volcanoes and river-like features exist on the surface.

In 1976 each of the Viking 1 and 2 probes placed landers on the surface of Mars. Both landers sent back a great deal of information about surface features and atmospheric conditions as well as conducting experiments to search for micro-organisms. No conclusive evidence of life was found.

The Mars Global Surveyor launched by the USA on 7 November 1996, consisted of an orbiter and robotic lander. The probe was inserted into a low altitude, nearly polar orbit on 12 September 1997 and it now circles Mars once every 2 h. The mission has studied the entire Martian surface, atmosphere and interior, and has returned more data about the red planet than all other missions combined.

Mars Pathfinder (USA) arrived at Mars on 4 July 1997. It used an innovative method of directly entering the atmosphere, assisted by a parachute to slow its descent and a giant system of airbags to cushion the impact. The landing site was an ancient, rocky, flood plain in Mars' northern hemisphere known as Ares Vallis. A six-wheel robotic rover, named Sojourner, rolled onto the Martian surface on 6 July. Mars Pathfinder returned 2.6 billion bits of data, including more than 16,000 images from the lander and 550 images from the rover, as well as more than 15 chemical analyses of rocks and extensive data on winds and other weather factors.

In April 2001, the United States launched the Mars Odyssey space probe. The probe carried instruments to analyse the chemical composition of the surface and rocks just below the surface. The mission also looked for the presence of water ice on Mars and looked for radiation hazards in the space surrounding Mars. Mars Odyssey went into orbit around Mars in October 2001.

In January 2004, NASA landed two robotic rovers (Spirit and Opportunity) on Mars, as part of the Mars Exploration Rover project. The rovers originally had 90-day missions that called for them to search for signs of past water activity on the red planet. Both rovers far outlasted their planned mission lengths. After becoming bogged in soft soil, Spirit was declared dead in 2011, but Opportunity has continued roving. Both rovers made big discoveries that have fundamentally reshaped scientists' understanding of Mars and its environmental history. One of the key discoveries was that water once existed on the Martian surface (Fig. 7.2).

Mars Express was Europe's first mission to another planet (launched in 2003, reaching Mars in 2005). It provided subsurface measurements with the first radar instrument ever flown to Mars, and discovered underground water-ice deposits. It sent back mineralogical evidence for the presence of liquid water throughout Martian history and measured the density of the planet's crust. The orbiter's unique orbit also has allowed it to make up-close studies of Phobos, the larger of Mars' two moons. The mission has been extended several times.

NASA's Mars Reconnaissance Orbiter (MRO), arrived at the red planet in March 2006 and spent half a year gradually adjusting

Fig. 7.2 The Mars exploration Rover named Opportunity has been exploring Mars since landing inside Eagle Crater on 25 January 2004. In its first
decade of driving on Mars, opportunity covered over 38 km (Credit: NASA).

the shape of its orbit. During 2007, the probe orbited the planet
once every 24 h and returned data to Earth at a rate faster than any
previous mission. The orbiter is examining Mars in unprecedented
detail including water and mineral distribution, features and
future landing sites. So far the probe has discovered channels in a
fossil delta, troughs in sand dunes and evidence that liquid or gas
has flowed through cracks in underground rocks. In December
2013 the orbiter imaged Curiosity and its tracks in Gale crater.

A US space probe named Phoenix was successfully launched
from Cape Canaveral, Florida, on 4 August 2007. The probe took
9 months to reach Mars. A robotic arm on the lander dug for clues
to past and present life in the polar region. Instruments checked
the soil for water and carbon-based chemicals (considered essential
for life). The lander completed its mission in August 2008, and
made a last brief communication with Earth on 2 November as

available solar power dropped with the Martian winter. The mission was declared concluded on 10 November 2008 after engineers were unable to re-contact the craft.

NASA launched Mars Science Laboratory (MSL) on 26 November 2011. The overall objectives include investigating Mars' habitability, studying its climate and geology, and collecting data for a manned mission to Mars. The mission successfully landed Curiosity, a Mars rover, in Gale Crater on 6 August 2012. Curiosity is about twice as long and five times as heavy as the Spirit and Opportunity rovers, and carries over ten times the mass of scientific instruments (see Figs. 7.3 and 7.4).

In November 2013, MAVEN (Mars Atmosphere and Volatile EvolutioN) was successfully launched, the first space probe devoted to understanding Mar's upper atmosphere. It entered orbit around Mars on September 22, 2014. The probe has already found that solar particles bury more deeply into the atmosphere than previously thought.

In 2016 NASA's Insight mission will place a lander on the planet. Insight will measure the seismic activity and internal heat flows on Mars. ESA and Russia are working on a two-pronged mission in 2016 and 2018, both looking for evidence of life past or present.

Position and Orbit

Mars orbits the Sun in an elliptical orbit that has the third highest eccentricity of all the planets' orbits. Its mean distance from the Sun is just over 228 million km, placing it about one and a half times further from the Sun than Earth. At perihelion, Mars is 208 million km from the Sun, while at aphelion it is 249 million km. The difference between the two is about 41 million km, whereas on Earth the difference between perihelion and aphelion is only 5 million km. This has a major influence on Mar's climate and results in a wide range of seasonal temperatures.

Mars orbits the Sun with a velocity of about 86,868 km/h and takes 687 Earth days to complete the trip. The planet takes 24.6 h (1.029 Earth days) to rotate once on its axis, which is tilted at an angle of 25.2° to the vertical.

Fig. 7.3 The Curiosity Rover on the surface of Mars (Credit: NASA).

About every 780 days Mars passes through a point in its orbit where it appears opposite the Sun in the sky (opposition). Because of its eccentric orbit, Mars distance at opposition varies, so its apparent size and brightness also change. The most favorable opposition is when Mars is closest to both the Earth and the Sun (this occurs about once every 17 years). Oppositions occurred on

Fig. 7.4 The Mars Rover Curiosity used its Mast Camera on 7 August 2014, to record this view of sedimentation in an ancient river bed (Credit: NASA/JPL-Caltech/MSSS).

Table 7.2 Significant space probes to Mars since 1996

Probe	Country of origin	Date launched	Notes
Mars Globe Surveyor	USA	1996	Mapped surface
Mars Pathfinder	USA	1996	Lander on surface
Mars Odyssey	USA	2001	Orbiter 3 years
Mars Explore Rovers	USA	2003	Two landers
Mars Express	USA	2003	Orbiter
Mars Recon. Orbiter	USA	2005	Orbiter
Phoenix	USA	2007	Lander
Mars Orbiter	India	2013	General stud
MAVEN	USA	2013	Study atmosphere

8 April 2014 and 22 May 2016. Mars can be studied easily from Earth using a telescope of moderate power. Dark markings and the white polar ice caps may be seen on the surface, depending on the distance Mars is from Earth.

Density and Composition

Even though Mars is more than half the diameter of Earth, it has only about one tenth of its mass. Details about the interior of Mars are limited because of the lack of seismic data.

The average density of Mars is the lowest of the terrestrial planets (3.95 g/cm^3 compared to Earth 5.52 g/cm^3). This suggests the iron-bearing core of Mars is smaller than Earth's core.

In fact the core is thought to have a radius of only 1100 km (but some estimates have it as high as 2000 km). The core makes up only about 6 % of the planet's mass, compared to Earth's core, which makes up about 32 % of its mass. A weak magnetic field suggests the core is no longer liquid or that currents within it are slow. Surrounding the core is a molten rocky mantle about 2200 km thick that is less dense than the core. The outer crust of the planet varies in thickness from about 20 km to 150 km (Fig. 7.5).

The strength of gravity on Mars is about a third less of Earth's gravity. A 75 kg person on Earth would weigh 735 N, but on Mars they would only weigh 270 N.

The Surface

Although Mars is much smaller than Earth, its surface area is about the same as the land surface area of Earth. Our first view of the surface of Mars was obtained from the Viking 1 lander in July 1976. Pictures revealed a rocky, desert-like terrain. Two weeks later Viking 2 set down on the opposite side of Mars, and images returned to Earth showed the same rocky surface as revealed by Viking 1. Both landing sites contained rocks ranging from pebbles to boulders in an orange-red, fine-grained soil.

Much of the Martian surface is very old and cratered, but there are also much younger rift valleys, ridges, hills and plains. There is a highly varied terrain on Mars with highlands in the southern hemisphere and lowlands in the northern hemisphere. The highlands are the oldest terrain (about 3 billion years old), and many parts are heavily cratered. The oldest terrain also contains

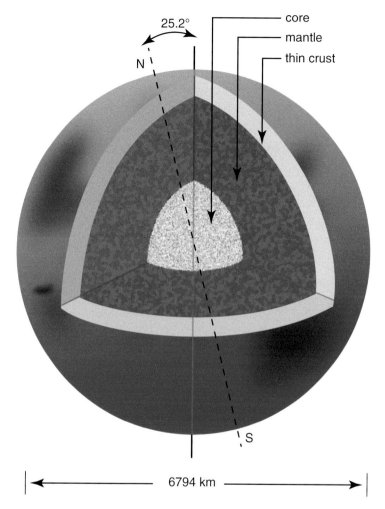

Fig. 7.5   The interior structure of Mars.

small channels that may have been carved by flowing water or dry ice. Smooth plains between the cratered areas are volcanic in origin.

The lowlands in the north contain mostly plains with few craters, indicating they probably formed after the period of bombardment by meteorites. The region between the highlands and lowlands is marked by an escarpment or long cliff. The reason for this abrupt elevation change could be due to a very large impact during Mar's past. A three-dimensional map of Mars that clearly shows these features was produced by the Mars Global Surveyor space probe.

The planet's western hemisphere contains a distinct bulge about 10 km high and 8000 km long, called the Tharsis Rise. This region contains the greatest concentration of volcanic and tectonic activity on Mars. Many volcanoes, fractures and ridges, and the enormous Valles Marineris canyon system, are linked to this rise. Valles Marineris was named after the Mariner 9 probe that discovered it. The canyon is about 8 km deep and 4500 km long. Smaller tributary canyons are as large as the Grand Canyon on Earth. Valles Marineris probably formed from rifting, or the pulling apart of the Martian crust, at the same time as the Tharsis Rise formed (see Figs. 7.1 and 7.6).

Fig. 7.6 A canyon in Valles Marineris on Mars filled with dense ground fog photographed from orbit by Mars Express (Credit: ESA).

Fig. 7.7 Olympus Mons is the largest volcanic mountain in the solar system (Credit: NASA).

Mars also contains the largest volcanic mountain in the solar system, Olympus Mons. This mountain rises to a height of 24 km above the surrounding plains, is more than 500 km wide and is rimmed by a cliff 6 km high. The volcano's summit has collapsed to form a volcanic crater or caldera about 90 km across (see Fig. 7.7).

There are three other prominent volcanoes on Mars: Arsia Mons, Pavonis Mons and Ascraeus Mons (see Fig. 7.1 left limb). Each is over 20 km high and forms part of a volcanic chain near the centre of Tharsis Rise. Alba Patera is a low-relief volcano about 2 km high and 700 km across situated near the rise's northern edge. Most of the volcanoes on Mars are in the northern hemisphere, while most of the impact craters are in the southern hemisphere.

The Martian surface contains many large basins formed when large meteors or asteroids have hit the surface. The largest basins formed by impacts are Hellas (with a diameter of about 2000 km), Isidis (1900 km) and Argyre (1200 km) (see Fig. 7.8).

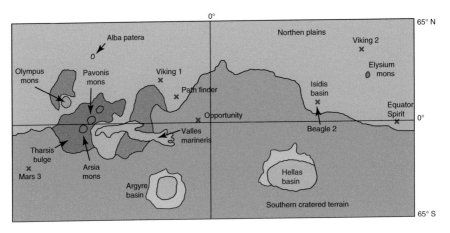

Fig. 7.8 Simplified map of geological features on Mars. The shaded area is the northern plain. The lower area is the southern cratered terrain.

The soil on Mars was analysed by the Viking landers and found to be slightly magnetic, indicating it contains iron. Further analysis showed the rocks at both landing sites were rich in iron, silicon and sulfur. As a result the Martian soil is described as an iron-rich clay. The soil is also rich in chemicals that effervesce (fizz) when moistened. Unstable chemicals called peroxides exist in the soil and these break down in the presence of water to release oxygen gas. Mars Pathfinder was able to identify the presence of conglomerates like those that are formed by running water on Earth. This evidence suggested Mars might have had a warmer past in which liquid water was stable.

Like Mercury and the Moon, Mars appears to lack active plate tectonics at present since there is no evidence of recent horizontal motion of the surface. Although there is no current volcanic activity, the Global Surveyor space probe showed that Mars might have had tectonic activity in its early history. Mars does not appear to have a crust made up of several large plates, like Earth has—it may be a single-plate planet.

Mars Global Surveyor found many landforms on Mars seem to have been formed or altered by running water. The channels that form valleys in the cratered highlands are similar to those formed by water on Earth. Some eroded valleys are huge, for example the Kasei Valles cuts over a kilometre deep into the volcanic plains of the Tharsis Rise, and is over 2000 km long. It is thought that some of the water flows in the past were huge in volume and occurred when internal heat or meteorite impacts released groundwater in sudden floods. Such flows were brief since Mars does not have enough water to sustain continuous flow. Some scientists believe that many of the geological features that appear to have been caused by liquid water, may have instead been formed by pyroclastic flows similar to those that occur during a volcanic eruption or by dry ice. In July 2014, NASA scientists using MRO data reported that gullies on Mars were likely to be formed by seasonal freezing of dry ice rather than liquid water.

If there is evidence of water in Mar's past, where is it today? Most of the water is believed to exist as permafrost in the northern lowlands and maybe underground in the heavily fractured and cratered highlands, and as ice at the poles. In December 2006, scientists comparing photographs taken in 1999 and 2005 from the Mars Global Surveyor's orbiting camera, discovered that water had flowed down the walls of a crater during this period.

Mars has two prominent polar ice caps, which can be seen through telescopes from Earth. The two polar regions of Mars are mostly covered with layered deposits of solid carbon dioxide (dry ice), with some dust and water ice. The mechanism responsible for layering is thought to be due to climatic changes. The Viking landers found that seasonal changes in the extent of the polar ice caps changes the atmospheric pressure by about 25 %.

During summer in the northern polar region, carbon dioxide returns to the atmosphere, leaving a cap of water ice. The southern polar region reduces its size during summer but stays as frozen carbon dioxide.

Life may exist in the permafrost or under the polar ice caps of Mars. On Earth, algae, bacteria, and fungi have been found living in ice-covered lakes in Antarctica, and so future explorations of Mars's ice may reveal life forms.

On 4 August 2011, NASA announced that MRO had found evidence of flowing salty water on the surface or subsurface of Mars.

In December 2013, scientists using Mars Express images, completed a topographical map of the Martian surface.

The Curiosity probe that landed in Gale crater found that rivers and streams once flowed over the crater floor, it also found

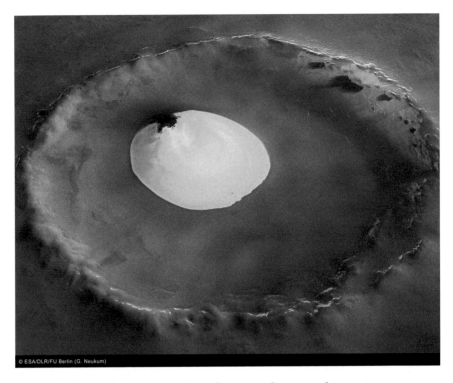

© ESA/DLR/FU Berlin (G. Neukum)

Fig. 7.9 A lake of water ice 200 m deep was discovered in an impact crater on Mars in 2005 by ESA's Mars Express. The crater is 35 km wide and has a maximum depth of 2 km (Credit: ESA).

evidence that a lake once existed in the region. A detector on Curiosity has made the first measurements ever of radiation on the surface of Mars. The detector found that Galactic cosmic rays and solar eruptions bombarded Mars, and their high-energy particles break the bonds that allow organisms to survive. The radiation would almost certainly be damaging to any microbial life on the surface and just below it. Many scientists on the Curiosity team believe that such radiation would damage the carbon compounds on Mars, and that this is a major reason why it has been so difficult to identify organics on the surface. In December 2014, Curiosity's instruments detected methane, the simplest organic compound, in both the atmosphere and surface of Mars. This indicates microbial life could live beneath the planet's surface.

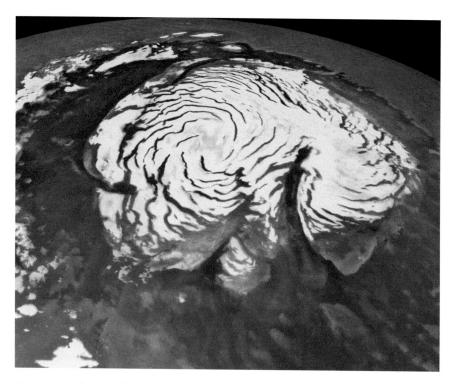

Fig. 7.10 The north polar cap of Mars has spiral shaped structures and a large canyon (Chasm Borale) that MRO radar images have shown to be caused by strong winds which blow from the *top* of the ice cap (Credit: NASA/MRO).

The Martian Atmosphere

The Martian atmosphere is very different from Earth's, but there are some similarities. Mars has a very thin atmosphere composed of 95.3 % carbon dioxide, 2.7 % nitrogen, 1.6 % argon, and less than 0.2 % oxygen. The atmosphere is near its saturation point with water vapour (0.03 %). In the past the Martian atmosphere may have been denser but it is now one-hundredth the density of Earth's atmosphere. This low value is partly due to the low gravitation field. Recent, isotopic studies of Mar's atmosphere for several elements, including hydrogen, argon, and carbon, suggest that the planet has lost between 25 % and 90 % of its original atmosphere. Data returned by the Phobos space probe suggests that the solar wind is carrying away the weakly held atmosphere at a rate of 45,000 tonnes per year.

Losing so much atmosphere would have left Mars much colder and drier, and liquid water would not have lasted on the surface. In 2014, the Maven space probe began to investigate the Martian atmosphere with a view to providing answers as to how the planet has been losing its atmosphere over time. Instruments on Maven detected comet dust in Mars' atmosphere and UV auroral glows (caused by particles from the solar wind). Maven also generated a map of a layer of ozone in the lower atmosphere of Mars and detected an ionosphere between 120 km and 480 km altitude.

The average air pressure on Mars is about seven-thousandths of Earth's, but it varies with altitude from almost nine-thousandths in the deepest basins to about one-thousandths at the top of Olympus Mons.

The minute traces of water vapour can at times form clouds, particularly in equatorial regions around midday. Early morning fogs also appear in canyons and basins. Temperatures around the poles are often low enough for carbon dioxide to form a thin layer of cloud.

The atmosphere is thick enough to support strong winds and dust storms that sometimes cover large areas of the planet. At times, the dust storms can hide surface features from Earth view. Such storms occur most often during the southern hemisphere's spring and summer. In 1971 Mariner 9's view of Mars was obscured by a dust storm that lasted for 2 weeks. In 1977, 35 dust storms were observed, and two of these developed into global storms. Global storms spread rapidly, eventually enshrouding the whole planet in a haze that can last a few months.

Mar's thin atmosphere produces a greenhouse effect but it is only enough to raise the surface temperature by 5°, which is much less than increases on Earth and Venus.

In December 2014, the Curiosity rover inside Mars' Gale crater, measured a ten-fold spike in methane in the atmosphere around it, and detected other organic molecules in a rock-powder sample collected by the rover's drill. This temporary increase is believed to be from some localized source.

Temperature and Seasons

The two Viking landers functioned as weather stations for two full Martian years. Their data, together with information from the orbiters, has given us a good picture of the weather on Mars.

Mars has a greater average distance from Earth and because of this it has a lower average surface temperature ($-60\,°C$). At perihelion, Mars receives about 45 % more solar radiation than at aphelion. As a result there is a large variation in surface temperatures during the Martian year. The coldest temperatures of $-125\,°C$ occur in winter at the south pole; this temperature is the freezing point of carbon dioxide. The warmest temperatures, of around $22\,°C$, occur during summer in southern mid-latitudes. The large difference between equatorial temperatures and polar temperatures produces a brisk westerly winds and low-pressure systems, similar to cyclonic systems on Earth.

Mars is tilted on its axis at 25.2°, which is similar to Earth's tilt of 23.5°, and so it experiences four seasons. Each season lasts about twice as long as Earth's because Mar's orbit is much larger and more elongated.

On Mars the Sun appears about half the size as it does on Earth.

In Mar's northern hemisphere, spring and summer are characterised by a clear atmosphere with little dust. White clouds may be seen at sunrise near the horizon and at higher elevations. During winter, falling temperatures around the northern polar ice cap cause carbon dioxide from the atmosphere to condense to renew the ice cap. The carbon dioxide ice comes and goes with the seasons, but a permanent ice cap of water ice remains.

The southern hemisphere summer occurs when Mar's is closest to the Sun and so southern summers are hotter than northern summers and winters are colder. During summer the southern polar ice-cap shrinks but a core of water ice remains.

Magnetic Field

Data from space probes indicates that Mars does not have an internal dynamo capable of generating a large global magnetic field. However, Mars may once have had such a dynamo. This is mainly supported by observations from the Mars Global Surveyor probe, which from 1997 to 2006 measured the magnetic field of Mars using a small magnetometer from an altitude of 100–400 km above the planet's surface. These measurements showed the existence of magnetic crustal fields on the planet's surface. However, the strength of the crustal magnetic field varies from place to place on the planet.

The weak magnetic field overall suggests that the core of Mars is no longer liquid or that currents in the core are slow. A computer model produced by scientists in 2009 suggests Mars's magnetic field may have been slowly weakened or knocked out by several large meteoroid impacts.

Mars also lacks a magnetosphere, which means its surface and atmosphere are exposed to attack by solar radiation. It is believed that fields detected on Mars are remnants of a magnetosphere that collapsed early in its history. The lack of a magnetosphere is thought to be one reason for Mars's thin atmosphere. Solar wind induced ejection of atmospheric atoms has been detected by probes orbiting Mars.

Martian Moons

Mars has two small moons that were discovered in 1877 by American astronomer Asaph Hall. Hall named them Phobos (meaning fear) and Deimos (meaning panic).

Phobos is the larger and innermost of the two moons being 21 km in diameter. It takes 7 h 39 min to orbit Mars, at a orbital radius of 9380 km. **Deimos** is only 12 km in diameter, takes 30 h 18 min to orbit Mars, at an orbital radius of 23,460 km. Both moons have an irregular shape, low density and many craters. They are hard to see from Earth because they are so small and reflect little light. The densities of each moon are so low they cannot be pure

Fig. 7.11 The two moons of Mars—Deimos and Phobos are both irregularly shaped and probably captured asteroids (Credit: NASA).

Table 7.3 Details of the moons of Mars

Name	Distance from Mars (km)	Period (days)	Diameter (km)	Discovered
Phobos	9380	0.32	21	1877
Deimos	23,460	1.26	12	1877

rock. Both moons orbit in nearly circular orbits (see Fig. 7.11 and Table 7.3).

Recent images from Mars Global Surveyor indicate Phobos is covered with a layer of fine dust about a metre thick. The Soviet spacecraft Phobos 2 detected a faint but steady out gassing from Phobos. Unfortunately, Phobos 2 failed before it could determine the nature of the material.

Phobos is slowly being pulled closer to Mars (1.8 m per century), and in about 50 million years it will either crash into the surface of Mars or break up into a ring. Deimos on the other hand appears to be getting further from Mars, slowing down as it does so.

Phobos has a potato shape and always has the same face turned towards Mars. If you were standing on Mar's equator you would see Phobos rising in the west, move across the sky in only 5½ h and set in the east, usually twice a day. Deimos takes about 2½ days to cross the Martian sky. Both moons are heavily cratered but Phobos has one large crater called Stickney (named after Angelina Stickney, Hall's wife) that is 10 km wide (this is visible on the right side of Phobos in Fig. 7.11). The grooves and streaks on the surface of Phobos were probably caused by the Stickney impact. Deimos is less cratered than Phobos. The largest crater on Deimos is only 2.3 km across.

It is thought that these two moons were asteroids captured by Mar's gravitational field. Evidence to support the asteroid theory is that both moons reflect very little of the light that falls on them and are very light for their size. The cratering on each moon suggests their surfaces are equally old—about 3 billion years. They are similar to C-type asteroids, which belong to the outermost part of the asteroid belt.

Further Information

For information about Mars and the various space missions check out
http://mars.jpl.nasa.gov/
http://www.nasa.gov/maven
www.space.com/mars/
For fact sheets on any of the planets including Mars check out
http://nssdc.gsfc.nasa.gov/planetary/planetfact.html

8. The Asteroid Belt

Highlights

- In February 2001, the NEAR probe landed on the asteroid Eros. A gamma-ray spectrometer on the probe was able to analyse material from the surface.
- The Japanese Hayabusa probe landed on the surface of the asteroid Itokawa on 9 May 2003. It collected surface material and returned it to Earth.
- The DAWN spacecraft has become the first spacecraft ever to orbit two worlds beyond Earth—Vesta in 2011/12 and Ceres in 2015.
- In January 2014, ESA scientists detected plumes of water vapour on Ceres.
- The largest asteroid, Ceres, was classified as a dwarf planet in 2006. A visit by the DAWN spacecraft in March 2015 captured some exciting images including unusual bright spots.

The asteroid belt is region between the orbits of Mars and Jupiter that contains about a million small rocky bodies, called asteroids. Most of these irregular shaped bodies are only a few kilometres in size, but they all orbit the Sun in much the same plane and direction. The belt contains about 200 asteroids larger than 100 km and about 750,000 up to 1 km in size. However, their combined mass is only about one-twentieth that of the Moon.

Individual asteroids are classified by their characteristic spectra, with the majority falling into three main groups: C-type, S-type, and M-type. These were named after and are generally identified with carbon-rich, stony, and metallic compositions, respectively.

There are some asteroids that have very elliptical orbits and cross Earth's orbit. These Earth-crossing asteroids have many craters on their surface because of impacts with other smaller

J. Wilkinson, *The Solar System in Close-Up*, Astronomers' Universe,
DOI 10.1007/978-3-319-27629-8_8,
© Springer International Publishing Switzerland 2016

bodies. Fragments of rock or iron ejected from the asteroids following impacts create bodies called meteoroids. A meteoroid entering Earth's atmosphere often heats up and glows, giving off light and appears like a like a shooting star or meteor across our night sky. Most meteoroids are about as old as the solar system itself.

The largest and first known asteroid, Ceres, is about 950 km in diameter. It contains about one-third the total mass of all the asteroids. In 2006, Ceres was classified as a dwarf planet because it orbits the Sun, has enough mass to form a spherical shape, has not cleared the area around its orbit and is not a satellite. The second largest asteroid is Vesta, with a diameter of around 550 km, closely followed by Pallas, an irregular shaped object about 540 km across. One of the smallest asteroids is 1991BA, discovered in 1991, and only 6 m across. The only asteroid that can be seen by the unaided eye is Vesta. It can be seen in dark skies when it is favorably positioned because it has a relatively high reflective surface.

Early Views About the Asteroids

The word 'asteroid' means 'star-like'. This name probably arose because, viewed from a telescope from Earth, asteroids look like a star.

Ancient observers on Earth did not know the asteroids because most cannot be seen with the unaided eye. Johann Elert Bode first suggested the idea that a planet-like body might exist between the orbits of Mars and Jupiter in 1768. In January 1801, the Sicilian astronomer Giuseppi Piazzi discovered a body in a position similar to that predicted by Bode. This body was called Ceres in honour of the Roman goddess of plants and harvest, and was the first asteroid to be discovered. In March 1802 the German astronomer, Heinrich Olbers discovered another faint asteroid that he called Pallas (after the Greek goddess of wisdom). Two more asteroids were discovered in the early 1800s were called Juno and Vesta.

By 1850, ten asteroids were known to orbit at average distances from the Sun of between 2.2 and 3.2 AU. These early findings were made by astronomers who spent many hours at the

Table 8.1 Details about the largest asteroid Ceres (a dwarf planet)

Distance from Sun	413,700,000 km (2.76 AU)
Diameter	950 km in diameter
Mass	9.46×10^{20} kg
Density	2.07 g/cm^3 or 2070 kg/m^3
Orbital eccentricity	0.08
Period of revolution	1681 Earth days or 4.60 Earth years
Rotation period	9.07 h
Orbital velocity	17.9 km/s
Axial tilt	3°
Average temperature	−106 °C
Atmosphere	Tenuous, some water vapour
Strength of gravity	0.27 N/kg

telescope observing changes in positions of celestial objects against the background of stars. In 1891 the German astronomer Max Wolf made the first photographic discovery of an asteroid. The object was named Brucia (the 323rd asteroid to be found). By 1923, the list of asteroids had grown to over a thousand.

During the early years of discovery, mythological names were given to the asteroids. All the early names were female (this naming scheme was later abandoned). Some asteroids were named after countries, for example, 1125 China, while others are named after scientists, for example, 2001 Einstein. Permanent numbers are assigned to asteroids once their orbits have been calculated and confirmed.

Astronomers are not certain how the asteroids originated, but many believe they are part of the solar nebula that failed to form a planet because of the strong gravitational pull of the nearby planet Jupiter. Others believe the asteroids are the remains of a planet that was pulled apart. This is unlikely since if all the asteroids were combined into a single planet, it would have a diameter of only 1500 km. Thus it is more likely that the asteroids are the remains of rocky bodies that have survived from the early solar system.

Probing the Asteroids

In October 1991, the NASA space probe Galileo took the first detailed photograph of an asteroid while en route to Jupiter. The asteroid, named **Gaspra**, is an irregularly shaped object measuring about 19 by 12 km in size. The Galileo probe also passed by the asteroid **Ida**, in August 1993. Both Gaspra and Ida are classified as S-type asteroids since they are composed of metal-rich silicates. Both asteroids are probably fragments of larger parent bodies that were broken apart by catastrophic collisions. Ida's surface is more heavily cratered than Gaspra's, but Ida is much older. Also, Ida, has its own companion, Dactyl, which is 1.5 km is diameter and orbits Ida at a distance of 100 km (see Figs. 8.1 and 8.2).

In 1996, NASA launched the **NEAR** (Near Earth Asteroid Rendezvous) space probe. This probe flew within 1216 km of the asteroid Mathilde in 1997. This encounter gave scientists the first close-up look of a carbon-rich C-type asteroid. This visit was unique because NEAR was not designed for fly-by encounters.

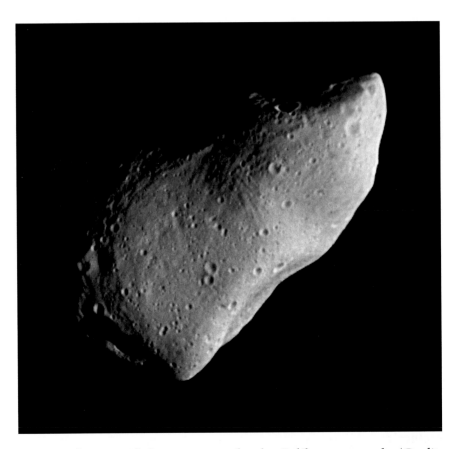

Fig. 8.1 The asteroid Gaspra as seen by the Galileo space probe (Credit: NASA).

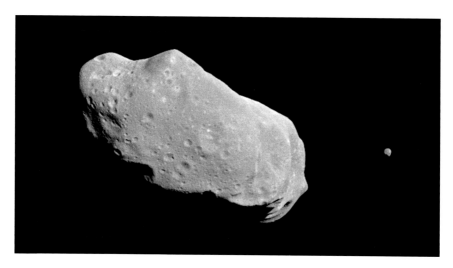

Fig. 8.2 The asteroid Ida and its companion Dactyl as seen by the Galileo probe (Credit: NASA).

The next year, NEAR flew past the asteroid Eros at a distance of 3829 km, and it went into orbit around Eros in February 2000. In March 2000, the probe was renamed NEAR-Shoemaker in honor of American astronomer Eugene Shoemaker, who had died not long before. In February 2001, NEAR-Shoemaker became the first spacecraft to land on an asteroid when it landed on Eros.

In July 1999 NASA's Deep Space 1 probe flew within 26 km of the asteroid Braille.

The Japan Aerospace Exploration Agency (JAXA) launched the **Hayabusa** probe on 9 May 2003. The probe's mission was to land on the surface of the asteroid Itokawa, and to collect samples and return them to Earth. The first attempt at landing failed but the second, in September 2005, was successful. Problems with the probe's engines delayed the return flight, however it eventually returned to Earth on 13 June 2010. The heat-shielded capsule made a parachute landing in the South Australian desert while the spacecraft broke up and incinerated in a large fireball. On 16 November 2010, JAXA confirmed that most of the particles found inside the Hayabusa sample container came from Itokawa. Japanese scientists found the samples were more similar to meteorites than known rocks from Earth, with concentrations of olivine and pyroxene. On 26 August 2011, scientists announced that the dust from Itokawa suggested it was probably part of a larger asteroid. The dust collected is believed to have been there for about 8 million years. Pictures from Hayabusa showed the surface of Itokowa is unlike any other solar system body yet photographed—a surface possibly devoid of craters.

The ESA's **Rosetta** probe passed within 3162 km of the asteroid 21 Lutetia in July 2010.

Lutetia has an irregular shape and is heavily cratered, with the largest impact crater reaching 45 km in diameter. The surface was found to be geologically heterogeneous and intersected by a system of grooves and scarps, which are thought to be fractures. It has a high average density, suggesting it is made of metal-rich rock (see Fig. 8.3).

In September 2007, NASA launched the **DAWN** spacecraft on a mission to the asteroid belt. Seeking clues about the birth of the solar system, the craft orbited Vesta between July 2011 and September 2012, before moving on to encounter Ceres in 2015. Both these asteroids are believed to have evolved more than 4.5 billion years ago, about the same time Mercury, Venus, Earth and Mars formed. Images taken by the Hubble Space telescope show these two asteroids are geologically diverse, but mysteries abound. The DAWN probe was the first spacecraft to go into orbit around a main belt asteroid, enabling a detailed and intensive study of the object (Table 8.2).

On 13 December 2012, China's lunar orbiter Change 2 flew within 3.2 km of the asteroid 4179 Toutatis on an extended mission. Around 2015 Japan Aerospace Exploration Agency plans to launch the improved Hayabusa 2 space probe and to return asteroid samples by 2020. Current target for the mission is the C-type asteroid 1999JU3. On 19 June 2014, NASA reported that asteroid 2011 MD was a prime candidate for captures by a robotic mission, perhaps in the early 2020s.

Position and Orbit

The asteroids orbit the Sun in a region between Mars and Jupiter between 2.0 and 3.5 AU from the Sun. The largest asteroid, **Ceres**, orbits at an average distance from the Sun of 413,700,000 km

Fig. 8.3 The ESA's Rosetta space probe took this image of the asteroid 21 Lutetia at closest approach in July 2010. The asteroid is 100 km in diameter and made of metal rich rock (Credit: ESA).

Table 8.2 Significant space probes to asteroids

Probe	Country of origin	Launched	Visited
Galileo	USA	1989	Gaspra 1991, Ida 1993
NEAR-Shoemaker	USA	1996	Mathilde 1997, Eros 2001
Deep space 1	USA	1998	Braille 1999
Hayabusa	Japan	2003	Itokawa 2011
Rosetta	ESA	2004	Lutetia 2010
DAWN	USA	2007	Vesta 2011, Ceres 2015

(2.76 AU). Ceres is spherical but is not to be considered a planet because it has not cleared its neighbourhood of other objects; instead it is classed as a dwarf planet. Its diameter is only one quarter the diameter of Earth's moon. Pallas orbits the Sun in a slightly elliptical orbit; it is dimmer than Ceres and has a diameter of only 540 km. Vesta is the brightest asteroid, orbiting at an average distance from the Sun of 2.36 AU.

Asteroids, which remain within the main asteroid belt, are referred to as main belt asteroids. In 1857 the American astronomer Daniel Kirkwood suggested that there would be gaps in the asteroid belt, created by gravitational perturbations of Jupiter. The existence of what became known as Kirkwood Gaps was confirmed in 1866. Such gaps were made through repeated alignments of asteroids with Jupiter. For example, an asteroid with an orbital period of exactly half that of Jupiter would, on every second orbit, be aligned with Jupiter.

The **Trojan asteroids** are two groups of asteroids, which travel around the Sun in the same orbit as Jupiter. These two groups are held in orbit by the combined gravitational forces of Jupiter and the Sun (see Fig. 8.4).

Some larger asteroids have been found to have smaller asteroid fragments orbiting them. The asteroid Toutatis consists of two similarly sized bodies orbiting each other. The asteroid Ida has a companion called Dactyl (see Fig. 8.2).

During 2007, astronomers announced that they have found strong evidence that sunlight can cause asteroids to spin more quickly by 'pushing' on the irregular surface features, and this accelerates or decelerates the rotation rate. The theory is that the Sun's heat serves as a propulsion engine on the irregular features of an asteroid's surface (Fig. 8.5).

Asteroid Collisions with Earth

Although there may be a million asteroids in the asteroid belt, their average separation is actually 10 million km, so collisions are not as common as one might expect.

On June 14, 1968, the asteroid Icarus passed within 6 million km of Earth. On 23 March 1989, the asteroid 1989 FC passed within 800,000 km of earth, and on 9 December 1994, asteroid 1994 XM1 passed within 105,000 km. This latter asteroid is only

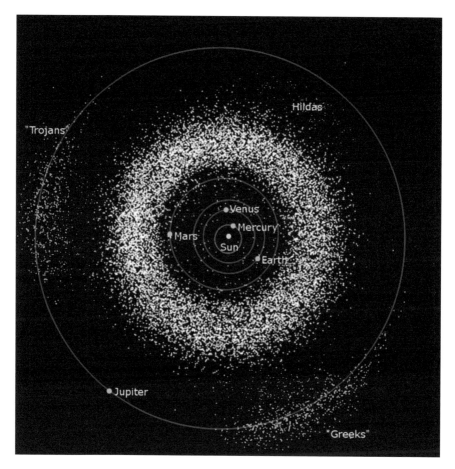

Fig. 8.4 The main belt of asteroids (*white*) is located between the orbits of the planets Mars and Jupiter. There are three other groups (Trojans, Hildas, Greeks) that share the orbit of Jupiter (Credit: NASA).

about 10 m in diameter but it could have done a lot of damage had it hit Earth.

Some asteroids have hit the Earth's surface and left craters. There is a 1-km-wide crater in northern Arizona that was made by an asteroid impact some 50,000 years ago. This asteroid was made of iron and nickel and was about 30 m wide. As recently as 1908, what is thought to have been a 100,000 tonne asteroid exploded in the Earth's atmosphere at an altitude of about 10 km with the force of a 20–30 megaton nuclear bomb. The explosion occurred over Siberia and the blast from the explosion flattened forests and burned an area about 80 km across. Many scientists believe an

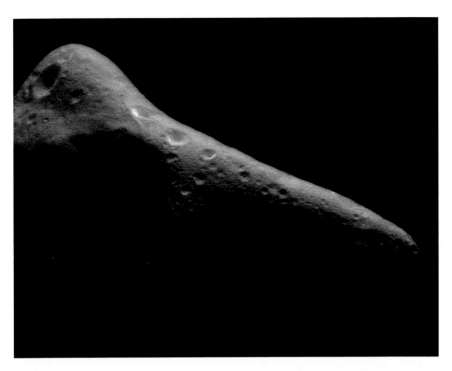

Fig. 8.5 The asteroid Eros is irregular in shape because of a series of collisions. It is 31 km long and its surface is covered by dust and rock fragments. This image was taken by the NEAR spacecraft in February 2000 from a distance of 289 km (Credit: NASA/NEAR).

asteroid impact about 65 million years ago was responsible for the extinction of the dinosaurs on Earth. The asteroid created a huge circular depression called the Chicxulub Basin centred in Mexico's Yucatan Peninsula. The diameter of the basin is about 180 km.

Size and Composition

Asteroids vary greatly in size. The largest asteroid, **Ceres**, is 950 km in diameter and contains about one third the total mass of all the asteroids. When it was discovered in 1801, Ceres was thought to be another planet because of its size. The discovery of other bodies in the same region made it an asteroid. Ceres was classified by the IAU in 2006 as a 'dwarf planet' because although

Fig. 8.6 An image of Ceres as seen by the Dawn probe on 19th February 2015 from a distance of 46,000 km. It shows a surface covered with small craters. Inside one crater are two bright spots thought to be due to water ice, volcanic action or salts reflecting sunlight. However, infrared images reveal the spots have different thermal properties (Credit: NASA).

it had a near spherical shape, it was too small to have cleared out the smaller chunks of matter in its orbital path (Figs. 8.6 and 8.7).

The second largest asteroid is **Vesta**. A recent analysis of Vesta's shape and gravity field using data gathered by the Dawn spacecraft has shown that Vesta is not spherical enough to be in hydrostatic equilibrium and is therefore not a dwarf planet. Vesta has several impact craters and a large concavity and protrusion near its south pole caused by impacts. The asteroid's crust is thicker than previously thought (about 80 km), and it lacks olivine on its surface a major component of planetary mantles). Vesta is the only asteroid that has an earth-like internal structure (core, mantle and crust). The asteroid also has sinuous gullies in some crater walls and a series of concentric troughs that are some of the longest chasms in the solar system. These troughs are thought to

Fig. 8.7 Ceres's northern hemisphere as seen by the Dawn probe on 15 April 2015 from a distance of 22,000 km (Credit: NASA/Dawn).

Table 8.3 Details about Vesta (the second largest asteroid)

Distance from Sun	353,260,000 km (2.50 AU)
Diameter	525 km in diameter
Mass	2.59×10^{20} kg
Density	3.456 g/cm^3 or 3456 kg/m^3
Orbital eccentricity	0.088
Period of revolution	1325 Earth days or 3.63 Earth years
Rotation period	5.34 h
Orbital velocity	19.3 km/s
Axial tilt	29°
Average temperature	−188 °C to −20 °C
Atmosphere	None
Strength of gravity	0.25 N/kg

be caused by shock waves after collisions (see Table 8.3 and Fig. 8.8).

A detailed geological map has been made of Vesta using data from the Dawn probe. The map shows Vesta is geologically divided into three provinces linked to the formation of its three largest

Fig. 8.8 Image of Vesta as seen by the Dawn spacecraft. The towering mountain at the bottom of the image (south pole) is more than twice the height of Mount Everest. The set of three craters known as the "snowman" can be seen at the *top left* (Credit: NASA/JPL-Caltech/UCLA/MPS/DLR/ IDA).

craters—the Veneneia basin, the Rheasilvia basin and the Marcia crater. The canyons around Vesta's equator are thought to be due to stress from the Rheasilvia impact (Fig. 8.9).

Scientists have found out about the composition of asteroids by analysing the light reflected from their surface and from analysing meteorite fragments. There are two main types of asteroids based on their composition. One group that dominates the outer part of the belt, is rich in carbon with a composition has not changed much since the solar system formed. The second group is located in the inner part of belt, and is rich in minerals such as iron and nickel.

Scientists believe Ceres and Vesta have followed a different evolutionary path. Vesta's origins seemed to have been hot and violent because it has basaltic flows on its surface. Vesta also

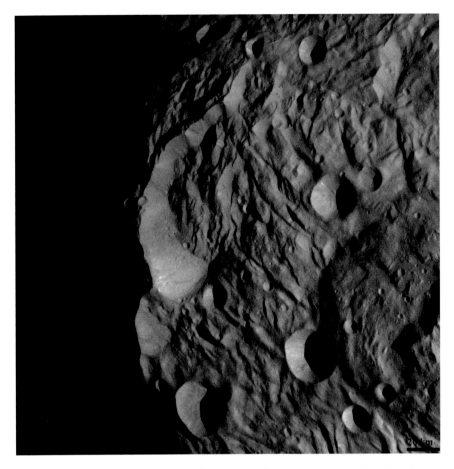

Fig. 8.9 South polar scarp on the asteroid Vesta, as seen by the Dawn spacecraft (Credit: NASA).

formed earlier than Ceres and is a very dry body. By studying these contrasts and comparing these two asteroids, scientists hope to develop a better understanding of the transition from the rocky inner regions of the solar system to the icy outer regions.

Asteroids become darker and redder with age due to space weathering. However evidence suggests most of the color change occurs rapidly, in the first hundred thousands years, limiting the usefulness of spectral measurement for determining the age of asteroids. Asteroids also contain traces of amino acids and other organic compounds, and some scientists speculate that asteroid impacts may have seeded the early Earth with the chemicals

Table 8.4 The ten largest Asteroids

Number	Name	Diameter (km)	Year discovered	Discovered by
1	Ceres	950	1801	G. Piazzi
4	Vesta	550	1807	H. Olbers
2	Pallas	540	1802	H. Olbers
10	Hygeia	443	1849	De Gasparis
704	Interamnia	338	1910	V. Cerulli
511	Davida	335	1903	R. Dugan
65	Cybele	311	–	–
52	Europa	291	1858	Goldschmidt
451	Patientia	281	–	–
31	Euphrosyne	270	–	–

necessary to initiate life, or may have even brought life itself to Earth (Table 8.4).

The Surface

Scientists have classified asteroids according to the amount of light they reflect. The dark stony kind reflects less than 5 % of the sunlight that falls on them. The brighter, light-coloured kind reflects about 20 % of incident light. Rarer third kinds resemble iron meteorites and may be the shattered core of older asteroids.

Photographs of most asteroids show they are covered with craters and dust, made by impact with other smaller rock-like bodies. The surfaces of asteroids are generally rock-like, with a layer of soil-like material.

On 12 February 2001, the NEAR probe landed on the asteroid Eros. Eros is irregular in shape and about 33 km long by 13 km wide. The first images of Eros showed it has an ancient surface covered with craters, grooves, house-sized boulders and other complex features. NEAR's gamma-ray spectrometer was able to analyse material in the surface to a depth of about 10 cm, detecting the elements iron, potassium, silicon and oxygen. The density of Eros is about 2.4 g/cm^3, about the same density as the Earth's crust. Now turned off, the NEAR probe could remain preserved in its present location, the vicinity of the huge, saddle-shaped feature called Himeros, for millions of years. As the asteroid orbits the Sun, the spacecraft's solar panels will be repeatedly turned toward the Sun, offering the possibility of reawakening the probe.

In 2012, NASA released the preliminary results of Dawn's study of Vesta. Vesta is thought to consist of a metallic iron–nickel core 214–226 km in diameter, an overlying rocky olivine mantle, with a surface crust. Data from Dawn suggested the dark spots and streaks on Vesta's surface were likely deposited by ancient asteroid impacts. The dark material contains hydrated minerals and might be carbon-rich. Gullies on the surface of Vesta are thought to have been eroded by transiently flowing liquid water. Numerous fragments of Vesta were ejected by collisions 1 and 2 billion years ago that left two enormous craters occupying much of Vesta's southern hemisphere. A recent analysis of Vesta's shape and gravity field using data gathered by the Dawn has shown that Vesta is currently not in hydrostatic equilibrium.

Observations of Ceres, the largest known asteroid, by NASA's Hubble Space Telescope, have revealed that the object may contain pure water beneath its surface. Scientists estimate that if Ceres were composed of 25 % water, it may have more water than all the fresh water on Earth. However, unlike the water on Earth, water on Ceres would be in the form of water ice and be located in the mantle. Today, Ceres' surface is too hot for ice to be stable anywhere except possibly, at the poles. If there ever was ice exposed at the surface, it has sublimed away. The shape of this asteroid is almost spherical, suggesting it may have an interior with a rocky inner core, and watery mantle with a thin, dusty outer crust.

Ultraviolet observations by spacecraft have revealed the existence of hydroxide water vapour near the north pole of Ceres. On 22 January 2014, ESA scientists using the far-infrared abilities of the Herschel Space Observatory reported the detection of water vapour on Ceres. Plumes of water vapour have been detected shooting up periodically from Ceres when portions of its icy surface warm slightly (such as when it is closer to the Sun).

The Atmosphere

The asteroids are much too small to retain an atmosphere. Any gases would be lost to space because of the very low gravitational attraction. Ceres may have a tenuous atmosphere containing water vapour.

Temperature

The maximum surface temperature of Ceres is –38 °C. This is relatively warm compared to the average surface temperature of a typical asteroid of –100 °C. This low temperature is largely because of the large distance between the Sun and the asteroid belt. The temperature on Vesta varies from –188 °C (dark side) to –18 °C (sunlit side). There are no seasons on the asteroids, although they do undergo day and night, dependent on which side is facing the Sun.

Magnetic Field

The NEAR probe that landed on the asteroid Eros also contained a magnetometer to measure any magnetic field. This instrument found no surface magnetic field exists.

It is unlikely that any asteroid would contain a magnetic field large enough to be detected.

Further Information

http://nssdc.gsfc.nasa.gov/planetary/
www.space.com/51-asteroids/
http://dawn.jpl.nasa.gov

9. Jupiter: The Gas Giant

Highlights

- Jupiter has a system of four thin rings composed of rocks and dust particles surrounding its atmosphere in an equatorial plane.
- New Horizon's is the fastest spacecraft ever (80,000 km/h) to travel between Earth and Jupiter taking only 13 months.
- NASA's Juno spacecraft, launched in August 2011, will be the first solar-powered spacecraft to orbit Jupiter.
- X-ray telescopes and the Hubble Space Telescope regularly detect auroras on Jupiter that are thousands of times more powerful than those on Earth. Auroras also occur on Jupiter's largest moon, Ganymede.
- The moon Io is the most volcanically active body in the solar system. It has nine giant erupting volcanoes on its surface and up to 200 smaller volcanoes.
- Jupiter's moon Europa contains water and molecular oxygen making it a likely candidate for life.

The planets Mercury, Venus, Earth and Mars are regarded as the inner planets of the solar system because they orbit close to the Sun. In contrast, the orbits of the four large planets—Jupiter, Saturn, Uranus and Neptune—are widely spaced at great distances from the Sun. The four inner planets are composed mainly of rock and metal, with surface features such as mountains, craters, canyons, and volcanoes. The outer planets on the other hand, rotate much faster and consist of vast, swirling gas clouds.

The gas planets do not have solid surfaces; their gaseous material simply gets denser with depth. The diameter of such planets is given for levels corresponding to a pressure of one atmosphere. What we see when looking at these planets is the tops of clouds high in their atmosphere.

J. Wilkinson, *The Solar System in Close-Up*, Astronomers' Universe,
DOI 10.1007/978-3-319-27629-8_9,
© Springer International Publishing Switzerland 2016

Jupiter is the first of the gas giants and the fifth planet from the Sun. This planet is the largest in the solar system and it travels around the Sun once every 11.86 years at an average distance of 780 million km. Jupiter is so large that over 1300 Earths could be packed into its volume. It is also twice as massive as all the other planets combined (318 times the mass of Earth). Jupiter also contains about 71 % of all the material in the solar system, excluding the Sun.

Jupiter formed from the same swirling mass of gas and dust as the Sun and other planets. But unlike the inner planets, Jupiter was far enough away from the Sun to retain its envelope of lighter gases, mainly hydrogen and helium. The outer layer of Jupiter forms a gaseous shell almost 20,000 km thick.

Astronomers have studied Jupiter for many years as it is the fourth brightest object in the sky (after the Sun, the Moon and Venus). Ancient observers knew the planet and its movement across the night sky had been accurately plotted against the background of stars for centuries. Because Jupiter takes about 12 years to orbit the Sun, it spends about a year in each constellation of the zodiac as seen from Earth. To the unaided eye, Jupiter appears as a brilliant white star-like object in the night sky.

Early Views About Jupiter

To the ancient Romans, Jupiter was the king of the gods, the ruler of Olympus and the patron of the Roman state. The planet was also associated with Marduk, the most important figure in Mesopotamian cosmology and the patron god of the city-state of Babylon. According to the story, Marduk fought with Tiamat, the goddess of chaos, and her 11 monsters. Marduk defeated them one by one and split Tiamat's body in two, thus dividing heaven from Earth. Marduk came to symbolize the rule of heavenly order over the universe. The wandering star Jupiter was placed in charge of the night sky.

With the invention of the telescope in 1608, the planet Jupiter could be studied in more detail from Earth. In 1610, Galileo observed Jupiter through his telescope and discovered its four largest moons, Io, Europa, Ganymede and Callisto. They are named after the mythical lovers and companions of the Greek

Table 9.1 Details of Jupiter

Distance from Sun	778,330,000 km (5.20 AU)
Diameter	142,984 km
Mass	1.90×10^{27} kg (318 times Earth's mass)
Density	1.33 g/cm^3 or 1330 kg/m^3
Orbital eccentricity	0.048
Period of revolution	4329 Earth days or 11.86 Earth years
Rotation period	9 h 50 min
Orbital velocity	47,016 km/h
Tilt of axis	3.12°
Average temperature	$-153\,°C$
Number of Moons	At least 63
Atmosphere	Hydrogen, helium
Strength of gravity	24.6 N/kg at surface

god Zeus. These moons became known as the Galilean moons after Galileo.

The motion of these moons around Jupiter provided evidence to support Copernicus's heliocentric theory of the motions of the planets. Galileo was arrested because of his support for the Copernican theory and he was imprisoned for the rest of his life.

In 1665 the French-Italian astronomer, Giovanni Cassini was the first person to see the Great Red Spot on Jupiter. Twenty-five years later he observed that the speeds of Jupiter's clouds vary with latitude. Nearer the poles, the rotation period of Jupiter's atmosphere is more than 5 min longer than at the equator (Table 9.1).

In January 2016 researchers at the British Museum announced that analysis of a newly found ancient tablet (dated 350–50 BC) has revealed that Babylonian astronomers had calculated the movements of the planet Jupiter using an early form of geometric calculus some 1400 years before the technique was used by the Europeans.

This means that these ancient astronomers had not only figured out how to predict Jupiter's paths more than 1000 years before the first telescopes existed, but they were using mathematical techniques that would form the foundations of modern calculus as we now know it. This tablet thus provides the key to understanding how the Babylonians used a trapezoid shape to predict Jupiter's position, which was integral to their beliefs about the weather, the price of goods, and the fluctuating river levels throughout the year.

Fig. 9.1 The planet Jupiter and the Great Red Spot as seen by the Wide Field Camera on the Hubble Space Telescope in April 2014. The Great Red Spot (*lower right*) is a high-pressure anticyclone in Jupiter's southern hemisphere (Credit: NASA and ESA teams).

Probing Jupiter

The first space probe to visit Jupiter was Pioneer 10 in December 1973. This was the first probe to venture out into the solar system beyond Mars. The probe returned over 20 low-resolution images of Jupiter's cloud system. A year later, Pioneer 11 returned 17 images during its closest approach to Jupiter; it then used Jupiter's strong gravitational pull to propel it towards Saturn. These two probes also recorded data of Jupiter's atmospheric temperature and pressure and took several pictures of its moons.

The probes recorded changes in Jupiter's atmosphere, particularly around the Great Red Spot and discovered Jupiter's huge magnetic field.

NASA's **Voyager 1 and 2** spacecraft flew by Jupiter in mid 1979 before proceeding on to Saturn. The two probes discovered that Jupiter has complicated atmospheric dynamics, lightning and auroras. These probes also found increased turbulence around the Great Red Spot. The winds to the north and south of the spot blow in opposite directions, seemingly fuelling the spot's rotation. Three new moons were discovered as well as a ring system.

In 1989, NASA launched the **Galileo** space probe from a space shuttle in Earth orbit. The probe was to rendezvous with Jupiter in 1995, after a trip that used a gravity assist from Venus. In July 1994, while still 225 million km from Jupiter, Galileo was able to observe fragments of the comet Shoemaker-Levy as they hit Jupiter. The fragments hit Jupiter at a speed of about 200,000 km/h over a period lasting about a week. The collisions left visible marks in Jupiter's atmosphere.

When Galileo reached Jupiter in 1995 it released a probe that descended into Jupiter's atmosphere about 150 km below the cloud tops. Data from this probe indicated that there is much less water than expected. Also surprising was the high temperature and density of the uppermost parts of the atmosphere. Recent observations by the Galileo orbiter suggest the probe may have entered the atmosphere at one of the warmest and least cloudy areas on Jupiter at that time.

As the **Ulysses** space probe passed by Jupiter in February 1992 it gathered data, which showed that the solar wind has a much greater effect of Jupiter's magnetic field than earlier measurements had suggested (see Table 9.2).

Table 9.2 Significant space probes sent to Jupiter

Probe	Country of origin	Launch year	Notes
Pioneer 10	USA	1972	Fly by in 1973
Pioneer 11	USA	1973	Fly by in 1974
Voyager 1	USA	1977	Fly by in 1979
Voyager 2	USA	1977	Fly by in 1979
Galileo	USA	1989	In orbit from 1995
Ulysses	USA/ESA	1990	Fly by 1992
New Horizons	USA	2006	Fly by in 2007
Juno	USA	2011	Orbit in 2016

In February 2007, the **New Horizons** probe flew by Jupiter. The reason for the fly by was to give it a gravitational boost, throwing it towards Pluto. This brief encounter was also used as a test run for both the spacecraft and its earthbound controllers in preparation for the Pluto encounter in 2015. The probe passed within 51,000 km of Jupiter. Images were taken of Jupiter's rings, its moon Io, and the Little Red Spot on Jupiter's surface. New Horizon's is the fastest spacecraft ever (80,000 km/h), having bridged the gap between Earth and Jupiter in only 13 months.

NASA's **Juno** spacecraft, launched in August 2011, will be the first solar-powered spacecraft to orbit Jupiter. The probe will enter a polar orbit in July 2016 and move around Jupiter observing its gravity and magnetic fields, weather and composition, and the connections between the interior, atmosphere and magnetosphere. The Europa Jupiter System Mission, due to launch around 2020, will engage in an extended study of the planet's moon system, particularly Europa and Ganymede, and settle the long-running scientific debate over whether an ocean of liquid water exists under Europa's icy surface.

Position and Orbit

The slightly elliptical orbit of Jupiter lies between the asteroid belt and Saturn. Jupiter has a mean distance from the Sun of just over 778 million km, placing it about 5.2 times farther from the Sun than is Earth. It travels around the Sun once every 11.86 years and it rotates on its axis with a period shorter than any other planet. The short rotational period has resulted in Jupiter becoming flattened or oblate. The equatorial diameter is 142,984 km, which is about 8000 km greater than its polar diameter. This shape suggests the interior is a liquid rather than a solid or gas. The planet's axis that is tilted at an angle of 3.12° to the vertical.

Density and Composition

Jupiter is more than 318 times more massive than the Earth and has twice as much mass as all the other planets combined. Jupiter emits about twice as much heat as it absorbs from the Sun. Its core temperature is estimated at 20,000 °C—about four times greater than Earth's core temperature. This heat is thought to be generated by the gravitational contraction of Jupiter by about 3 cm per year. The core pressure may be about 100 million times greater than on Earth's surface.

The shape of the planet and its strong gravitational field suggest Jupiter must have a dense core about 10–20 times the mass of Earth. However, overall, Jupiter's average density is lower than that of Earth—only 1.33 g/cm^3 compared to Earth's 5.52 g/cm^3.

Jupiter is about 90 % hydrogen and 10 % helium with traces of methane, water, ammonia and 'rock'. This composition is very close to the composition of the nebula that the solar system formed.

There are three main regions to Jupiter's interior. The outer layer of Jupiter that we see from Earth is the top of the outer layer of clouds. The Galileo probe found that these clouds are mostly gaseous molecular hydrogen and helium. With increasing depth and hence pressure, the gases become more like liquids. The hydrogen becomes crushed into a liquid form called metallic hydrogen. Metallic hydrogen's high electrical conductivity and the rapid rotation of the planet, give rise to Jupiter's intense magnetic field and radiation belts. Most of the interior of Jupiter (its mantle) is therefore mostly liquid metallic hydrogen. Below the mantle, the third region or core of Jupiter is thought to consist of rocky material mainly iron and silicates. The core is about 20,000 km in diameter (Fig. 9.2).

The strength of gravity on Jupiter is 2.5 times that of Earth's gravity. This means that a 75 kg person who weighs 735 N on Earth would weigh 1845 N on Jupiter.

The Surface

The outer layers of Jupiter form a shell, mostly of gaseous hydrogen. Although this layer is about 20,000 km thick, there is no solid surface. The gas just gets thicker and thicker, until the pressure is three million times the air pressure at sea level on Earth. At this point, hydrogen becomes crushed into liquid metallic hydrogen.

Seen from Earth, Jupiter is one of the brightest planets in the sky. Viewed through a telescope, its disc is crossed by numerous belts or zones of various colours including red, orange, brown and

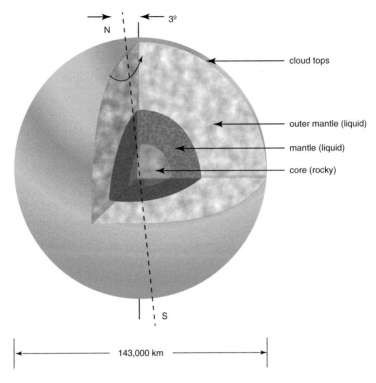

cloud tops

outer mantle (liquid)

mantle (liquid)

core (rocky)

143,000 km

Fig. 9.2 Interior structure of Jupiter.

yellow. The brighter zones are regions where fluids from within the planet are rising to the surface to cool, while the darker belts are regions where material is descending.

Also visible on the surface of Jupiter is the Great Red Spot first observed by Giovanni Cassini through his telescope in 1665. This spot is a huge, oval-shaped atmospheric feature located in the southern hemisphere. The size of the spot varies but is roughly 30,000 km in length and 12,000 km in width. The Pioneer and Voyager missions suggested that the spot is a hurricane-like storm whose red colour may be caused by the presence of red phosphorus and yellow sulfur in the ammonia crystals. Infrared observations show the spot is a high-pressure region whose cloud tops are much higher and colder than the surrounding regions. It is not fully understood how the spot lasts for so long, as it must be absorbing a lot of energy to keep surviving. The edge of the spot rotates at a speed of about 360 km/h. The spot moves east to west around Jupiter but stays about the same distance from the equator (see Fig. 9.1).

The Atmosphere

Spectra analysis of Jupiter's atmosphere shows it is 86 % by mass hydrogen and 13 % helium. The remainder consists of small amounts of simple compounds such as methane, ammonia, and water vapour.

Data from the Galileo probe released into Jupiter's atmosphere only goes down to about 150 km below the cloud tops. It is thought that the high pressure and radiation on Jupiter destroyed the Galileo probes sensors.

Three distinct layers of clouds are thought to exist on Jupiter. The upper layer is the coldest (-153 °C) and contains mainly ammonia ice crystals. The middle layer contains crystals of ammonium hydrosulfide and a mixture of ammonia and hydrogen sulfide. The lower layer contains water ice. The vivid colours seen in Jupiter's clouds are probably due to chemical reactions between the elements in the atmosphere. The colours seem to correlate with altitude: blue the lowest clouds, followed by browns and whites, with reds highest.

Clouds in Jupiter's turbulent atmosphere move at high velocity in east-west belts parallel to the equator. The winds blow in opposite directions in adjacent belts. Data from the Galileo probe indicate the winds travel at about 600 km/h and extend down thousands of kilometres into the interior. The winds are mainly driven by Jupiter's internal heat and the planet's rapid rotation.

Pictures from the Pioneer and Voyager probes showed changes often occur in Jupiter's atmosphere. The most notable of these was in the region around the Great Red Spot. At the time of the Pioneer probes (1973–1974) the spot was surrounded by a white zone. By 1979 when Voyager visited Jupiter, a dark belt had crossed the spot, and there was increased turbulence around the area. Measurements showed the spot rotates anti-clockwise over a period of about 6 days. The winds north of the spot blow in the opposite direction to those south of the spot. Lightning 10,000 times more powerful than any seen on Earth has also been detected in Jupiter's atmosphere.

Other oval shaped features called eddies or circular winds can be seen in the atmosphere of Jupiter. These eddies move about

Fig. 9.3 Close up image of Jupiter and the Great Red Spot as seen by the Hubble Space telescope. The moon Ganymede is also shown (*lower right*) (Credit: NASA/ESA).

within the zones in which they are trapped by opposing winds. They usually appear white in colour. The Great Red Spot is a huge eddy (see Fig. 9.3).

Jupiter's Ring System

Jupiter also has a system of four rings surrounding its atmosphere in an equatorial plane. These rings are much fainter and lighter than those around Saturn and can't be seen from Earth through normal telescopes. The rings were discovered by the Voyager probes in 1979, but they have since been imaged in the infrared from ground based telescopes, the Hubble Space Telescope (HST) and by the Galileo probe. Observations of the rings require the largest available telescopes.

The ring system's four main components are: a thick inner torus of particles known as the "halo ring"; a relatively bright,

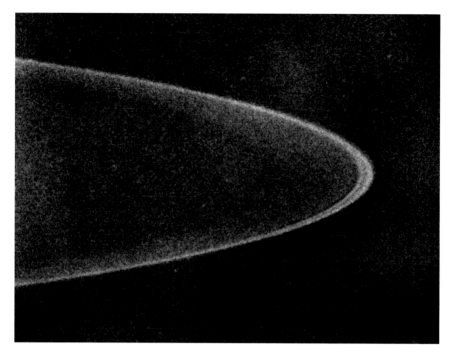

Fig. 9.4 The Long Range Reconnaissance Imager on the New Horizons probe snapped this photo of Jupiter's ring system as it flew by in 2007. This image shows a narrow ring, about 1000 km wide, with a fainter sheet of material inside it (caused by fine dust that diffuses in toward Jupiter) (Credit: NASA, GSFC, NSSDC).

exceptionally thin "main ring"; and two wide, thick and faint outer "gossamer rings", named for the moons of whose material they are composed: Amalthea and Thebe. The main ring is about 6500 km wide and 30 km thick. Its inner edge is about 123,000 km above Jupiter's cloud tops (see Fig. 9.4).

The rings are dark and are composed of very fine-grained dust particles and rock fragments, and unlike Saturn's rings, they do not contain any ice. Galileo found evidence that the particles are continuously being kicked out of orbit by radiation from Jupiter and the Sun. High-resolution images obtained in February and March 2007 by the New Horizons spacecraft revealed a rich fine structure in the main ring. The rings are probably re-supplied with material by dust particles formed by micrometer impacts on the four inner moons. The New Horizons spacecraft conducted a deep search for new small moons inside the main ring. While no satellites larger than 0.5 km were found, the cameras of the space-craft detected seven small clumps of ring particles. Spectra of the main ring obtained by the HST, Keck, Galileo and Cassini have shown that particles forming it are red in colour.

Temperature and Seasons

The temperature near the cloud tops of Jupiter measures about −153 °C. Temperatures increase with depth below the clouds, reaching 20 °C at a level where the atmospheric pressure is about ten times as great as it is on Earth. At a depth of about 20,000 km below the cloud tops the temperature is about 10,000 °C. Below this depth, the pressure and temperature are high enough to transform liquid hydrogen into liquid metallic hydrogen. The pressure at Jupiter's centre is about 80 million atmospheres and the tempera-ture about 24,000 °C, which is hotter than the surface of the Sun.

Because Jupiter takes 11.86 years to orbit the Sun and it has a small axial tilt, there are not any real seasons on Jupiter.

Magnetic Field

Jupiter has a huge magnetic field, about 20 times stronger than Earth's magnetic field. Its magnetosphere extends only a few mil-lion kilometres towards the Sun, but away from the Sun it extends to a distance of more than 650 million km (past the orbit of Saturn). Many of the moons of Jupiter lie within its magnetosphere. The magnetic field contains high levels of radiation and energetic particles that would be fatal to space travellers. The Galileo probe discovered a new intense radiation belt between Jupiter's rings and the upper most atmospheric layers. This belt is about ten times stronger as Earth's Van Allen radiation belts, and contains high-energy helium ions (see Fig. 9.5).

Jupiter emits radio waves strong enough to be picked up by radio telescopes on Earth. Scientists use these waves to calculate the rotational speed of Jupiter. The strength of the waves varies as the planet rotates and is influenced by Jupiter's magnetic field. The radio waves come in two forms. The first is a continuous emission from Jupiter's surface; the second is a strong burst that occurs

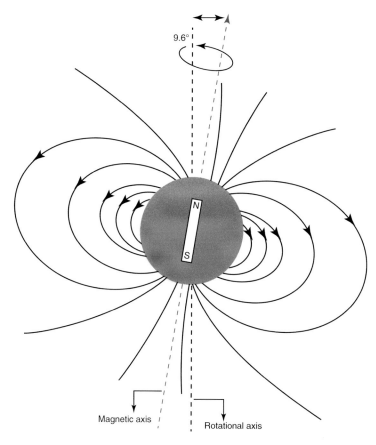

9.6°

N

S

Magnetic axis

Rotational axis

Fig. 9.5 Jupiter's magnetic poles are offset from the rotational axis by nearly 10°.

when the moon Io passes through certain regions of the magnetic field and radiation belt.

X-ray telescopes and the Hubble Space Telescope regularly detect auroras on Jupiter. These auroras however are thousands of times more powerful than those on Earth. On Earth, the most intense auroras are caused by outbursts of charged particles from the Sun interacting with the polar magnetic field of Earth. On Jupiter, however, the particles seem to come from the moon Io that has volcanoes that spew out oxygen and sulfur ions. Jupiter's strong magnetic field produces about 10 million volts around its poles, and this field captures the charged particles and slams them into the planet's atmosphere. The particles interact with

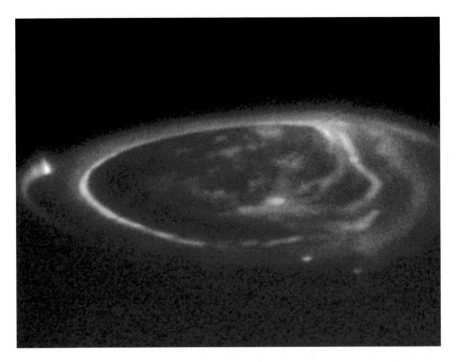

Fig. 9.6 An aurora around Jupiter's north pole. Imaged by the Hubble Space Telescope in UV light (Credit: NASA/HST).

molecules in the atmosphere and the result is intense X-ray auroras, virtually all the time (see Fig. 9.6).

Moons of Jupiter

At least 63 natural satellites or moons orbit Jupiter. Galileo was the first to observe the four largest moons of Jupiter through his telescope in 1610. These moons, now known as the Galilean moons, orbit Jupiter in an equatorial plane, creating the appearance of a mini-solar system. In order of distance from Jupiter, the four Galilean moons are Io, Europa, Ganymede, and Callisto (Table 9.3).

The first three Galilean moons, are locked together into a 1:2:4 orbital resonance by tidal forces from Jupiter. In a few hundred million years, Callisto will be locked in too, orbiting at exactly twice the period of Ganymede and eight times the period of Io (Fig. 9.7).

Table 9.3 Details of the Galilean moons of Jupiter

Name of moon	Distance from Jupiter (km)	Period (days)	Diameter (km)	Discovery year
Io	420,700	1.77	3630	1610
Europa	671,000	3.55	3138	1610
Ganymede	1,070,000	7.16	5262	1610
Callisto	1,883,000	16.69	4800	1610

Fig. 9.7 The four Galilean moons of Jupiter (Credit: NASA).

Io is about the same size as Earth's Moon and it orbits around Jupiter once every 1.77 days. This moon is the most volcanically active body in the solar system. Images from Voyager 1 showed Io has nine giant erupting volcanoes on its surface and up to 200 smaller volcanoes. These sulfurous eruptions give Io a white-, yellow- and orange-coloured surface.

Io's volcanoes are all relatively flat. The largest volcano, called Pele, is about 1400 km across. There are also mountains up to 10 km high, but these are not volcanic. Voyager also discovered numerous black spots scattered across Io, which are thought to be volcanic vents through which eruptions occur. The volcanic eruptions change rapidly. In the 4 months between the arrivals of Voyager 1 and Voyager 2 some of the eruptions had stopped, while others had begun, and deposits around the vents also changed visibly.

Io is thought to be volcanically active because of huge tidal forces created by Jupiter. Jupiter's gravity pulls the surface of Io so much that the surface flexes, or bends back and forth. This movement generates enough heat to melt the interior and produce Io's hot spot volcanism. Io's thin atmosphere is mostly sulfur dioxide gas produced by the volcanoes. At night, some of this gas freezes, and produces the white areas seen on the surface. Io has little or no water (see Fig. 9.8).

Europa is the second of the Galilean moons and the fourth largest moon of Jupiter. It is slightly smaller than Earth's moon

Fig. 9.8 A huge 100 km-high volcanic plume (eruption) is seen rising above the surface of Jupiter's moon Io. Lower down in the image, close to the shadow line, is a second, ring-shaped plume—about 75 km high (Credit: NASA).

and orbits Jupiter once every 3.5 days. Europa is locked by gravity to its planet so that the same side always faces toward Jupiter. Io and Europa have a similar composition, consisting of mainly silicate rock. Unlike Io, Europa has a thin outer layer of ice and a layered internal structure, probably with a metallic core. Its surface is relatively smooth with no mountains and very few craters. Some astronomers think topographical features and a layer of liquid water may exist below the ice-covered surface. If Europa's ocean is proven to exist, it would possess more than twice as much water as Earth. Fractures in the surface of this moon may be due to tidal forces caused by Jupiter.

In 1995, astronomers discovered the atmosphere of Europa is very thin and contains molecular oxygen. This oxygen is thought to be generated by sunlight and charged particles hitting Europa's icy surface producing water vapour that is split into hydrogen and oxygen. The hydrogen escapes, leaving behind the oxygen (see Fig. 9.9).

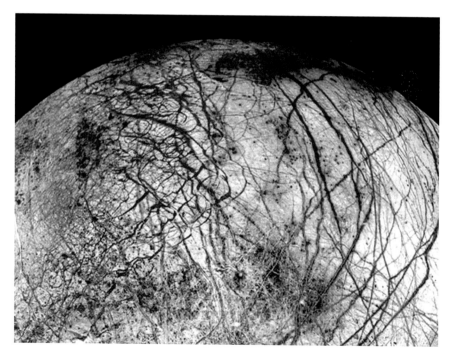

Fig. 9.9 Europa's icy surface is covered by numerous streaks and cracks thought to be caused by tidal forces from Jupiter (Credit: NASA).

In 2014 researchers announced they have clear visual evidence of Europa's icy crust expanding.

The surface of Europa is riddled with cracks and ridges. Surface blocks are known to have shifted in the same way blocks of Earth's outer ground layer on either side of the San Andreas fault move past each in California. Many parts of Europa's surface show evidence of extension, where miles-wide bands formed as the surface ripped apart and fresh icy material from the underlying shell moved into the newly created gap—a process similar to seafloor spreading on Earth.

Ganymede is the third of the Galilean moons and is the largest moon in the Solar System. Its diameter is larger than the planet Mercury, although it's mass and density is much less. Ganymede orbits Jupiter in synchronous rotation once every 7.16 days at a distance of about 1 million km.

Ganymede has both dark and light areas on its surface. The dark areas are old and heavily cratered. The lighter regions are young, have few impact craters but do contain many grooves and ridges. The largest feature on Ganymede is Galileo Regio, a dark circular area of ancient crust 4000 km in diameter, which contains an abundance of craters.

The crust is thought to be about 75 km thick and contain an outer layer of ice. Beneath lies a mantle of either water or ice, and a rocky or silicate-rich core. The Hubble Space Telescope has recently found evidence of oxygen in Ganymede's atmosphere, very similar to Europa's atmosphere.

The Galileo space probe found that Ganymede has its own magnetic field embedded inside Jupiter's huge field. This is probably generated in a similar fashion to the Earth's, as a result of moving, conducting material in the interior. The magnetic field of Ganymede causes aurorae, which are ribbons of glowing, hot electrified gas, in regions circling the north and south poles of the moon. When Jupiter's magnetic field changes, the aurorae on Ganymede also change, "rocking back and forth". By watching the rocking motion of the aurorae, scientists working with the Hubble Space Telescope announced in 2015 that they believe a large amount of saltwater exists beneath Ganymedes's crust. The subterranean ocean is thought to have more water than all the water on Earth's surface. A deep ocean under the icy crust of Ganymede opens up further exciting possibilities for life beyond Earth.

Callisto is the fourth of the Galilean moons and the second largest moon orbiting Jupiter. It is only slightly smaller than the planet Mercury but has only one-third its mass. Callisto orbits Jupiter once every 16.69 days in synchronous rotation at a distance of about 2 million km.

Callisto's surface is a dark, ancient and icy crust, covered with many old impact craters. In fact it is the most heavily cratered object in the solar system. The craters and impact basins are relatively flat because of the nature of the surface. The largest impact basin, Valhalla, is about 3000 km in diameter and is surrounded by bright concentric rings of fractured ice. Valhalla may have formed as a result of a huge asteroid impact. Numerous smaller craters cover this feature, which suggests an age of about 4 billion years.

Unlike Ganymede, Callisto has no grooved terrain, suggesting little if any tectonic activity occurred. It probably cooled very rapidly. Voyager instruments measured a temperature range of $-118\,°C$ during daytime and $-193\,°C$ at night.

Data from the Galileo probe suggests Callisto has little internal structure, and is composed of about 40 % ice and 60 % rocky iron. Its atmosphere is very tenuous and contains mainly carbon dioxide. There is no evidence of a magnetic field.

Other Moons of Jupiter

In addition to the four Galilean moons, Jupiter has at least 59 other moons. The non-Galilean moons tend to be irregular in shape and smaller than 185 km in diameter. Four of these moons are closer to Jupiter than Io, the rest orbit in regions beyond Callisto. The outer bodies are probably captured asteroids, while the inner ones are probably pieces broken off a larger body.

The four inner moons are Metis, Adrastea, Amalthea and Thebe. The largest of these is **Amalthea**, which was discovered in 1892 by the American astronomer Edward Barnard, and is only 167 km across. Amalthea was originally thought to be innermost moon, but the Voyager probes found Metis and Adrastea to be

Table 9.4 Details of the inner moons of Jupiter

Name	Distance from Jupiter (km)	Period (days)	Diameter (km)	Discovered (year)
Metis	127,690	0.29	43	1979
Adrastea	128,690	0.30	16	1979
Amalthea	181,400	0.50	167	1892
Thebe	221,900	0.68	98	1979

closer to Jupiter. Amalthea is a dark, heavily cratered irregularly shaped body with a reddish colour. The moons Metis, Adrastea and Thebe were discovered by the Voyager 1 probe in 1979 (Table 9.4).

The four inner moons and the Galilean moons, all orbit Jupiter in an equatorial plane in near-circular orbits. The outer moon's have more elliptical orbits and are more inclined to Jupiter's orbital plane. More surprisingly, all the moons after Carpo orbit in the opposite direction to that of the other satellites, and opposite to the direction of Jupiter's spin. These characteristics together with their small size suggest these outer satellites are captured asteroids, not originally part of the Jupiter system.

As of 2007, Jupiter has four small inner moons, four large Galilean moons, and 55 other tiny satellites of sizes ranging from 1 km to 170 km. The tiny 'asteroid-like' satellites orbit between 7,284,000 km and 30,290,000 km distance from the planet and may be better called 'moonlets'.

Further Information

For fact sheets on any of the planets including Jupiter check out
http://nssdc.gsfc.nasa.gov/planetary/planetfact.html
www.space.com/jupiter/
www.nasm.si.edu/etp/
www.nasa.gov/juno

10. Saturn: The Ringed Planet

Highlights

- Space probes have revealed that Saturn's rings are actually composed of hundreds of narrow, closely spaced 'ringlets'. Some moons actually orbit inside the rings.
- The moons of Saturn are numerous and diverse ranging from tiny moonlets less than 1 km across to the enormous Titan which is larger than the planet Mercury.
- Radar images taken by the Cassini spacecraft of Saturn's largest moon, Titan, have shown large lakes of hydrocarbons.
- In June 2005, the Cassini spacecraft detected auroral emissions associated with the solar wind around Saturn's poles.
- The Cassini spacecraft has found evidence of water spewing from geysers on the moon Enceladus.
- Titan, Saturn's largest moon, is the only moon in the solar system to have clouds and a dense atmosphere.
- In 2014 astronomers reported that Titan has a subsurface ocean made of water mixed with ammonia, and that the ocean may be as salty as the Dead Sea on Earth.

Saturn is the second of the gas giants and the sixth planet from the Sun. This planet is the second largest in the Solar System with a diameter of 120,536 km. It travels around the Sun once every 29.46 years at an average distance of 1430 million km. At its closest approach to the earth, Saturn is about 1,278,000 km away.

Saturn travels around the Sun in an elliptical orbit. Its distance from the Sun varies from about 1509 million km at its farthest point to about 1350 million km at its closest point.

Saturn is about 85 % the size of Jupiter but twice as far from Earth. It is over 95 times as massive as the Earth and, with the exception of Jupiter, has more mass than all the other planets combined. However, Saturn has the lowest density of all the

J. Wilkinson, *The Solar System in Close-Up*, Astronomers' Universe,
DOI 10.1007/978-3-319-27629-8_10,
© Springer International Publishing Switzerland 2016

Table 10.1 Details of Saturn

Distance from Sun	1,430,000,000 km (9.54 AU)
Diameter	120,536 km
Mass	5.68×10^{26} kg (95.18 times Earth's mass)
Density	0.69 g/cm^3 or 690 kg/m^3
Orbital eccentricity	0.056
Period of revolution	10,768 Earth days or 29.46 Earth years
Rotation period	10 h 40 min
Length of year	29.5 Earth years
Orbital velocity	34,704 km/h
Tilt of axis	26.73°
Average temperature	−185 °C
Number of moons	62
Atmosphere	Hydrogen, helium
Strength of gravity	10.4 N/kg at surface

planets, only 0.69 g/cm^3, which is less than the density of water and roughly half the density of Jupiter. This low density means Saturn must be composed of light elements.

Saturn formed from the same swirling mass of gas and dust as the Sun and other planets. Like Jupiter, and unlike the inner planets, Saturn was far enough away from the Sun to retain its envelope of lighter gases, mainly hydrogen and helium. As it orbits the Sun, Saturn spins on its axis, at a rapid rate. Its axis is titled at an angle of about 27° from the perpendicular.

Astronomers have studied Saturn for many years, as it is easily visible in the night sky of Earth. Ancient observers knew it, and its movement across the night sky has been accurately plotted against the background of stars for centuries. To the unaided eye, Saturn appears as a brilliant yellow-orange star like object in the night sky. Saturn's main feature is its spectacular ring system, which can be seen through a telescope from Earth. The only other planets to have rings are Jupiter, Neptune and Uranus, but their rings aren't as prominent as Saturn's rings and they can't be easily seen through a telescope from Earth (Table 10.1).

Early Views About Saturn

Saturn has been observed in the night sky since prehistoric times. Mesopotamian astronomers called Saturn the 'the old sheep' or 'the eldest old sheep', while the Assyrians described the planet as a sparkle in the night sky and named it 'Star of Ninib'. To the

Fig. 10.1 Image of Saturn and its rings. Credit: NASA.

ancient Romans, Saturn was the god of agriculture, while the Greeks called it Cronus, after Zeus's father, the overthrown ruler of the universe. Cronus was the also son of Uranus and Gaia. Saturn is the root of the English word 'Saturday'.

The modern era for Saturn began in 1610 when Galileo first observed it through his telescope, and described it as a triple-bodied object. Other observers thought Saturn had 'handles' or 'ears'. In 1659, Christiaan Huygens reported that Saturn was circled by a broad, flat ring and had a moon; this was to be called Titan. In 1676, Italian astronomer, Giovanni Cassini, discovered a gap in Saturn's ring system. With modern telescopes, Earth based astronomers have found Saturn has two prominent rings (A and B) and two faint inner rings (D and C). The gap between the A and B rings is now known as the Cassini division. A much fainter gap dividing the A ring and F ring is known as the Encke division after German astronomer Johann Franz Encke who allegedly saw it in 1838. Pictures taken by the Voyager probes show two additional faint outer rings G and E.

Probing Saturn

People on Earth have observed Saturn through telescopes based on Earth and in space since 1610 when Galileo first observed it, and more recently through space based telescopes.

The first space probe to visit Saturn was Pioneer 11, on 1 September 1979. The probe passed within 21,000 km of the planet and within 3500 km of its outer ring. It travelled under the ring system and sent back many useful pictures of the rings. However, the images showed little new information about Saturn's clouds and atmosphere.

In November 1980, Voyager 1 passed within 124,123 km of Saturn before moving out of the solar system. Voyager 2 also encountered Saturn on August 26, 1981, getting to within 101,335 km of the planet before proceeding on to Uranus and Neptune. The Voyager probes provided many pictures and data about Saturn. They found three new moons around Saturn, four additional faint rings and provided pictures of atmospheric circulation. Before the Voyager probes, information about Saturn's atmosphere was limited because astronomers could see only the tops of the clouds from Earth. The Voyager probes identified long-lived oval shaped structures inside the clouds and revealed three layers of clouds with slightly different compositions.

In 1994, the Hubble Space Telescope, while orbiting Earth at 28,000 km/h, captured the first images of aurora in Saturn's atmosphere. It also captured images of topographical features on Saturn's largest moon, Titan, which suggest a continent once existed on this moon.

The **Cassini mission** to Saturn (named after Giovanni Cassini) was one of the most ambitious ever attempted. It was joint venture of NASA, the European Space Agency (ESA) and the Italian Space Agency (known as ASI for its acronym in Italian), and was designed to explore the whole Saturnian system, the planet itself, its atmosphere, rings and magnetosphere and some of its moons. Launched in 1997, the Cassini space probe reached Saturn in 2004, and went into orbit around the planet. Cassini plunged between Saturn's two outer rings at 80,000 km/h before it slowed down enough to be captured by Saturn's gravity and begin its 4-year orbit of the planet. Instruments on board Cassini detected an eruption of atomic oxygen in Saturn's E-ring.

Cassini has also taken pictures of Saturn's largest moon, Titan. In July 2004, Titan was found to be surrounded by a thick atmosphere, with areas of water ice on its surface. Cassini also released a smaller probe called Huygens on 24 December 2004. Twenty days later, the probe entered Titan's atmosphere at about 6 km/s, and landed via parachute on Titan's surface, on 14 January 2005. The probe landed in mud like wet clay covered by a thin crust. The first images showed a pale orange, rock-strewn, eroded landscape with drain channels. The ground temperature was a chilling −180 °C. Huygen's was the first successful attempt by humans to land a probe on another world in the outer Solar System.

By the end of 2007, Cassini had flown by Titan 40 times and mapped over 60 % of Titan's 'Lake District' north of latitude 60°. Dark areas on the surface are believed to be filled with a mixture of liquid ethane, methane and dissolved nitrogen. Some of the lakes appear to be fed by rivers that flow down from the surrounding hill country to shorelines of bays, peninsulas and islands. The rivers have many tributaries among the uplands. A few lakebeds appear dry (Fig. 10.4).

From 2004 to November 2, 2009, the Cassini probe discovered and confirmed eight new moons of Saturn. Its primary mission ended in 2008 when the spacecraft had completed 74 orbits around the planet. The probe's mission was extended to September 2010 and then extended again to 2017, to study a full period of Saturn's seasons. In April 2013 Cassini sent back images of a hurricane at the planet's north pole 20 times larger than those found on Earth, with winds faster than 530 km/h (Table 10.2).

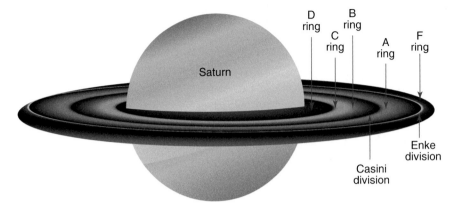

Fig. 10.2 Structure of Saturn's rings.

Position and Orbit

Saturn is the second largest planetary member of the solar system and is the sixth planet from the Sun. Its orbit is slightly elliptical and lies between Jupiter and Uranus. Saturn has a mean distance from the Sun of just over 1430 million km, placing it about 9.5 times farther from the Sun than Earth. It travels around the Sun once every 29.46 years and it rotates on its axis with a period of 10 h 40 min. As with Jupiter, the short rotational period has resulted in Saturn becoming flattened or oblate. The equatorial diameter is 120,536 km, which is 10 % more than its polar diameter of 108,728 km. This shape suggests the interior is a liquid rather than a solid or gas.

Density and Composition

Saturn's is the least dense of all the planets in the solar system, mainly because of its composition of light gases. Like Jupiter, Saturn is a gaseous planet composed of about 75 % hydrogen and 25 % helium with traces of water, methane, ammonia and 'rock' similar to the composition of the primordial solar nebula from which the solar system was formed. Saturn's mass is about 95 times greater than Earth's but it has 800 times the volume.

Fig. 10.3 Image of Saturn and its largest moon Titan as seen by the Cassini spacecraft. The shadow of Saturn's rings can be seen on the surface (Credit: NASA, ESA).

The interior of Saturn is also similar to Jupiter in that is contains a rocky core, a liquid metallic hydrogen mantle and a liquid outer layer of molecular hydrogen. Traces of various ices are also present. With increasing height, the outer layer of liquid hydrogen becomes gaseous.

Because of its smaller mass and size, Saturn's interior is less compressed than Jupiter's. The core is about 25,000 km in diameter, while the mantle is about 12,000 km thick. Saturn's core contains around 26 % of the total mass of the planet, as opposed to around 4 % for Jupiter. The temperature, pressure and density inside the planet all rise steadily toward the core, which, in the deeper layers of the planet, cause hydrogen to transition into a metal (Fig. 10.5).

As with Jupiter, Saturn's interior is hot, about 12,000 °C at the core, and the planet radiates 2.5 times more energy into space than it receives from the Sun.

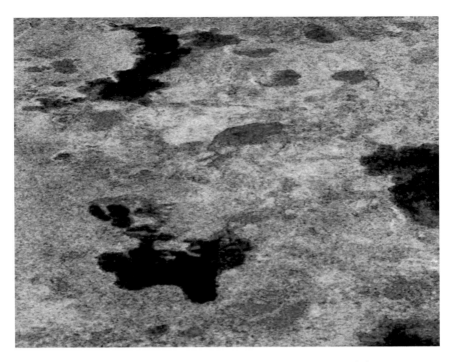

Fig. 10.4 Radar image taken by Cassini of the hydrocarbon lakes on Titan, Saturn's largest moon (Credit: NASA, ESA).

Table 10.2 Significant space probes to Saturn

Probe	Country of origin	Launched	Comments
Pioneer 11	USA	1973	Fly by in 1979
Voyager 1	USA	1977	Fly by in 1980
Voyager 2	USA	1977	Fly by in 1981
Cassini	USA, ESA, ASI	1997	In orbit from 2004

The strength of gravity on Saturn is only slightly more than Earth's gravity (10.4 compared to 9.8 N/kg). This means that a 75 kg person weighs 735 N on Earth, but on Saturn would weigh 780 N.

The Surface

Saturn's surface and interior are similar to those of Jupiter. Saturn's mantle is surrounded by ordinary liquid hydrogen, so there is no solid surface layer. When observing Saturn, we are

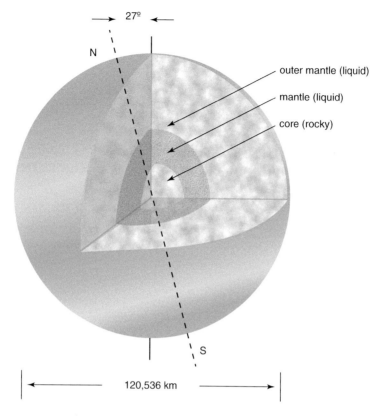

Fig. 10.5 Internal structure of Saturn.

looking at its cloud tops. These cloud tops lack the colours visible on Jupiter. Photographs taken by Voyager 1 show faint bands across Saturn's surface but these are nowhere near as prominent as those on Jupiter. This is mainly because, the gravitational pull of Saturn is much weaker than Jupiter's, and hence the layers of gases are more weakly held together. The banded appearance of the cloud layer is thought to be caused by differences in the temperature and altitude of the atmospheric gas masses.

Saturn does have some long-lived spots like the Great Red Spot on Jupiter. In 1990, the Hubble Space Telescope observed an enormous white cloud near Saturn's equator that was not present during the Voyager encounters. In November 1994, another spot was observed near Saturn's equator. This storm-like spot was 12,700 km across which is about the same size as Earth.

The Atmosphere

Saturn's atmosphere contains mostly hydrogen (96 %) and helium (3 %). Trace amounts of ammonia, acetylene, ethane, propane, phosphine and methane have been detected in Saturn's atmosphere. Ultraviolet radiation from the Sun causes methane photolysis in the upper atmosphere, leading to a series of hydrocarbon chemical reactions with the resulting products being carried downward by eddies and diffusion. This photochemical cycle is modulated by Saturn's annual seasonal cycle.

The atmosphere consists of a banded pattern, similar to Jupiter's, but the bands are fainter and wider near the equator. There are three layers of clouds. The lower layer of clouds contains water ice crystals. The middle layer contains clouds of ammonium hydrosulfide, while the uppermost layer contains ammonia ice crystals. The clouds generally rotate with Saturn with a period of 10 h 14 min at the equator to 10 h 40 min at high latitudes. The clouds appear yellow in colour and move in zones parallel to the equator, with winds that alternate from east to west between zones. Wind speeds are generally higher than those on Jupiter. This high-velocity wind of 1800 km/h has remained fairly constant over decades. Saturn also has storms like those seen on Jupiter, but they are less visible and less frequent, although they last longer.

Because Saturn is 9.53 times further from the Sun than Earth, its atmosphere receives only 1 % of the solar energy Earth receives. It radiates more than twice this amount from its interior. Gases being heated by the interior and Saturn's fast rotation generate circulation patterns in the atmosphere.

A strange hexagonal wave pattern around the north polar vortex in the atmosphere of Saturn was first noted in the Voyager images. The sides of the hexagon are each about 13,800 km long, which is longer than the diameter of the Earth. The structure rotates with a period of 10 h 40 min (the same period as that of the planet's radio emissions) which is assumed to be equal to the period of rotation of Saturn's interior. The hexagonal feature does not shift in longitude like the other clouds in the visible atmosphere. Most astronomers believe the pattern was caused by some standing-wave pattern in the atmosphere.

Imaging of the south polar region of Saturn by the Hubble Space telescope indicates the presence of a jet stream, but no strong polar vortex nor any hexagonal standing wave.

The Rings

The most prominent feature of Saturn is its ring system, which encircles the planet around its equator. The rings do not touch Saturn. As Saturn orbits the Sun, the rings tilt at the same angle as the equator. Sometimes we see the rings edge on from Earth and sometimes they are nearly upright. The Voyager space probes showed much more detail about the rings than could be seen from Earth, and four additional rings were discovered, bringing the total to seven.

Closer examination of the rings by space probes has revealed that the seven rings are actually composed of hundreds of narrow, closely spaced 'ringlets'.

The closest ring to Saturn is the faint D ring. The inner edge of this ring lies about 6700 km from the cloud tops. The C ring begins at about 14,200 km altitude. The densest of the rings is the B ring that begins at about 31,700 km above the cloud tops. The Cassini Division, which is a gap about 4700 km in width, is at an altitude of 57,200 km. The gap probably formed as a result of the gravitational pull between ring particles and Saturn's moon Mimas. Beyond this gap, the A ring begins at 76,500 km altitude, followed by the narrow, faint F ring. Beyond the F ring is the tenuous G ring, discovered by the Voyager space probe, and the even more tenuous E ring (the outermost ring). The bulk of the ring system spans about 275,000 km but is only about 1.5 km thick.

Saturn's rings, unlike the rings of other planets, are very bright because they reflect light. The rings consist of countless small particles, ranging in size from a centimeter to 10 m across. In order to measure the size of the particles in Saturn's rings, scientists measured the brightness of the rings from many angles as the spacecraft flew around the planet. They also measured changes in radio signals received as the craft passed behind the rings. If all the particles were compressed to form a single body, the body would be only about 100 km across. Data from the Voyager spacecraft confirmed that the particles consist of ice and ice-coated rocks.

Voyager also identified two tiny satellites, Prometheus and Pandora, each measuring about 50 km across, orbiting Saturn on either side of the F ring. They help to keep the icy particles into a well defined, narrow band about 100 km wide. The F ring also contains ringlets that are sometimes braided and sometimes separate. Dark radial 'spokes' that appear and disappear in the B ring are thought to be caused by Saturn's magnetic field.

Observations made by the Cassini probe in 2006 showed that the D ring isn't flat like the other rings. It appears to have corrugations like a tin roof. These corrugations are thought to have been caused by an impact as recently as 1984.

Saturn's rings are thought to have formed from a cloud of particles that came from the breakup of a moon or from material that did not combine to form a moon. Moons that orbit within the rings act as shepherd satellites to create sharp edges and gaps between the rings.

Temperature and Seasons

At the cloud tops and in the rings, the temperature is about −185 °C. Frozen water is in no danger of melting or evaporating at these cold temperatures. Temperature and pressure increases with depth below the cloud tops. In the outer core, the temperature reaches about 12,000 °C and the pressure about 12 million times the pressure on Earth's surface.

Because Saturn is titled on its axis and it takes 29.5 years to orbit the Sun, any season on Saturn would last more than 7 Earth years.

Magnetic Field

Saturn's magnetic field was first detected with the fly-by of NASA's Pioneer 11 spacecraft in 1979. Convection currents in the mantle of liquid metallic hydrogen generate the strong magnetic field. Saturn's field is about 36 times less powerful than the field of Jupiter but 570 times more powerful than Earth's field. Because the magnetic field is less powerful than Jupiter's fewer charged particles are trapped in Saturn's magnetic field. The rings and moons also absorb some charged particles.

Saturn's magnetosphere is intermediate in size between Earth and Jupiter's, but it extends beyond the orbit of Saturn's moon Titan. Data from space probes show that Saturn's magnetosphere contains radiation belts similar to those of Earth. Variations in the magnetic field are thought to be responsible for the presence of dark spokes seen moving in Saturn's rings.

On Earth, the magnetic polar axis and the rotational axis vary by about 11°, but on Saturn the two axes are within 1° of each other (see Fig. 10.6).

The Cassini spacecraft has detected auroral emissions around both poles of Saturn. The blue-ultraviolet emissions are thought to be caused by hydrogen gas being excited by electron bombardment. The images showed that the auroral light responded rapidly to changes in the solar wind strength. In 2013, astronomers using the Hubble Space Telescope captured new images of the dancing auroral lights at Saturn's north pole. The ultraviolet images capture moments when Saturn's magnetic field is affected by bursts of particles streaming from the Sun. It appears that when particles from the Sun hit Saturn, the planet's magnetotail collapses and later reconfigures itself, an event that is reflected in the dynamics of its auroras.

Moons of Saturn

Saturn has 62 moons, nearly as many as Jupiter. The moons fall into three groups—there are 21 moons between 133,000 km and 527,000 km from the planet, three between 1,221,000 km and

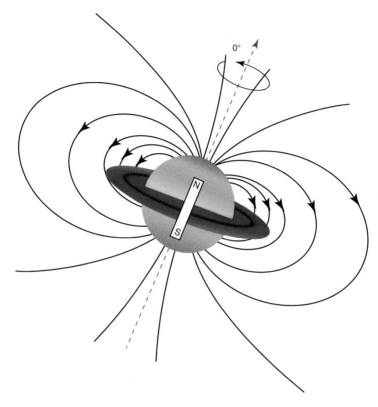

Fig. 10.6 Saturn's magnetic field. Shown are the magnetic axis and rotational axis.

3,560,000 km, and the rest between 11,294,000 km and 24,500,000 km.

The moons range in diameter from about 3 km to 5150 km but the smaller ones are more asteroid-like and may not be true moons. Some astronomers call them 'moonlets'. Most of these moons are icy worlds heavily covered with craters caused by impacts very long ago. Some astronomers believe that the moons may have condensed from a series of gas rings cast off from Saturn about 4.5 billion years ago.

There were 18 known moons orbiting Saturn when the Cassini spacecraft began its historic journey to the planet in 1997. During Cassini's 7-year journey to Saturn, Earth-based telescopes discovered 13 additional moons. When Cassini reached Saturn in 2004, three more moons were discovered (Methone,

Table 10.3 Largest moons of Saturn (in order of increasing distance from Saturn)

Name	Distance from Saturn (km)	Period (days)	Diameter (km)	Year discovered
Mimas	185,520	0.94	397	1789
Enceladus	238,000	1.37	500	1789
Tethys	294,000	1.89	1060	1684
Dione	377,400	2.74	1120	1684
Rhea	527,100	4.52	1530	1672
Titan	1,221,850	15.95	5150	1655
Iapetus	3,560,800	79.33	1440	1671

Pallene and Polydeuces). On 1 May 2005, Cassini found another moon, hidden in a gap in Saturn's outer A ring.

In 2006, astronomers using the 8.2 m Subaru telescope in Hawaii detected nine more moons orbiting Saturn. These moons or satellites are about 6–8 km in size and they travel on highly eccentric, retrograde orbits (opposite to the planet's rotation). These objects were probably captured by Saturn's gravity.

As of mid-2008, a total of 62 moons or satellites have been detected around Saturn.

Many of the moons have been officially named, the rest have been given temporary numbers until they are fully confirmed.

Only seven of the known moons of Saturn are massive enough to have collapsed into a spherical shape (see Table 10.3). The rest are irregular in shape suggesting they are captured asteroids. Unlike Jupiter, Saturn has only one planet-sized moon; this is called **Titan**. Titan is second in size only to Ganymede among the moons in the solar system, and it is also larger than the planet Mercury. It can be seen fairly easily through telescopes from Earth.

Christiaan Huygens discovered Titan in 1655, the same year he discovered the rings. It has a diameter of 5150 km and it orbits Saturn at a distance of about 1.2 million km. Titan takes about 16 days to orbit Saturn and it also takes this time to rotate once on its axis. Thus this moon always has the same side facing Saturn.

Titan is thought to be made of half water ice and half rock or silicates. A mantle of ice and an icy crust that may contain some liquid water surrounds its rocky core. Titan is the only moon in the solar system to have a dense atmosphere. Brown-orange clouds in the atmosphere completely obscure its surface and little sunlight reaches the surface. Voyager data showed most of the atmosphere is

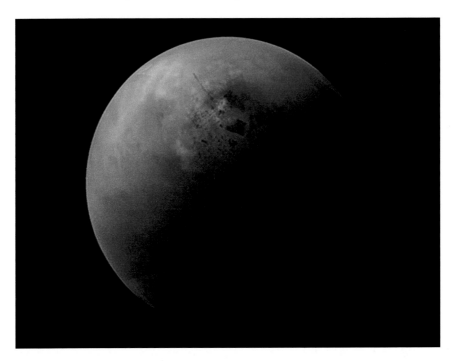

Fig. 10.7 Saturn's largest moon is Titan—it is the only moon with a substantial atmosphere. In 2004 the Huygens probe took this photo of the surface of Titan while descending through Titan's atmosphere (Credit: NASA).

nitrogen gas (94 %) with the rest mainly methane (5 %). The nitrogen gas may have originally been in the form of ammonia, which broke up into hydrogen and nitrogen. Hydrogen, being light, may have escaped Titan's weak gravity. There are traces of hydrocarbons such as, methane, acetylene, ethane, ethylene, and propane. All the oxygen is present as water ice. The atmosphere is four times as dense as Earth's but because of the weak gravity, atmospheric pressure is only 1.6 times greater than Earth's (Fig. 10.7).

The temperature on Titan is about −178 °C, which is below the freezing point of water, but near the freezing point of methane. Thus it is expected that the surface contain lakes of hydrocarbon liquids like ethane. Nitrogen reacts with these hydrocarbons to produce other compounds, some of which are the building blocks of organic molecules essential for life.

Pictures taken by the Cassini spacecraft in July 2004, showed a murky landscape with a variety of features, such as giant equatorial sand dunes, polar lakes, and methane-soaked mud flats. So far, only one mountain ridge has been detected. There is evidence of volcanoes, flows and calderas. The northern region contains well-defined lakes, channels and islands. The first infra-red pictures revealed water ice as dark patches and masses of clouds in the southern hemisphere.

Cassini also mapped interaction between the huge magneto-sphere that surrounds the Saturn system, and Titan's dynamic atmosphere. The 80,000 km wide gas cloud that follows Titan as it orbits the planet is evidence that its atmosphere is breaking up.

On 14 January 2005, the Huygens probe (released from Cassini) entered Titan's atmosphere at about 6 km/s, and its heat shield reached 8000 °C. Three parachutes were used to slow the probe down and it landed on the surface with a 'splat' in what appeared to be mud. The images of the surface showed a pale orange, eroded landscape of rocks and ice blocks, together with what looked like drainage channels. A thin crust gives the surface a squishy consistency. Titan also has large plains containing longitudinal hydrocarbon sand dunes that run for hundreds of kilometres. The dunes are up to a kilometre wide and as high as 300 m.

On 23 June 2014, NASA reported that it had strong evidence that nitrogen in the atmosphere of Titan came from materials associated with comets in the Oort cloud, and not from the materials that formed Saturn in earlier times.

Scientists now believe that Titan has a subsurface ocean made of water mixed with ammonia. On 2 July 2014, NASA reported the ocean inside Titan maybe as salty as the Dead Sea on Earth.

Other Moons of Saturn

Excluding Titan, the six largest moons range in diameter from 390 km to 1530 km. In order of increasing distance from Saturn, these moons are Mimas, Enceladus, Tethys, Dione, Rhea and Iapetus. These moons have average densities around 1.2 g/cm^3, which suggests that they are made mostly of water ice with some rock. Having formed in a cold environment, these bodies retained water, methane, ammonia and nitrogen that condensed from the solar nebula. Some astronomers believe these mid-sized moons condensed from gas rings that surrounded Saturn about 4500 million ago.

Mimas has a diameter of 392 km and is 185,520 km from Saturn. It takes only 23 h to orbit Saturn and is difficult to observe from Earth because it is so close to Saturn. Mimas has the distinction of resembling the Death Star from Star Wars Episode IV. The Voyager probe showed the surface of Mimas is dominated by a large impact crater, called Herschel, which is 130 km across (one third the diameter of Mimas). The walls of this crater are about 5 km high and parts of its floor measure 10 km deep. It has a central peak that rises 6 km above the crater floor. Fractures can be seen on the opposite side of Mimas that may be due to the large impact. Mimas has one of the heaviest cratered surfaces in the solar system, but they are all much smaller than Herschel (see Fig. 10.8).

Enceladus has a diameter of 500 km and is 238,020 km from Saturn. The surface of this moon is covered with a smooth layer of water ice that makes it the most reflective of any known planetary body. There are many craters in one hemisphere and very few in the other hemisphere. The young surface of this moon was seen by Voyager 2 to contain a number of different types of formations, including ice flows, faults and striations. The crust is probably thin and lying on top of a molten interior (see Fig. 10.9).

During 2006, the Cassini spacecraft had found evidence of icy water particles spewing from geysers on Enceladus. As a result some scientists have now placed this moon on the short list of places most likely to have extraterrestrial life. High-resolution images snapped by the orbiting Cassini probe confirmed the eruption of icy jets and giant water plumes from geysers at Enceladus' south pole. If any life exists on the moon, it probably would be in the form of microbes or other primitive organisms. Icy particles ejected from Enceledus are thought to have created Saturn's outermost E ring.

Tethys has a diameter of 1060 km and is 294,660 km from Saturn. The Italian astronomer Giovanni Cassini discovered it in 1684. Like Mimas, the surface of this moon is heavily cratered in places, but there are also some smooth areas. The largest impact crater is Odysseus with a diameter of 400 km (making it larger than the moon Mimas). This crater is very old with a flat floor that corresponds to the curvature of the moon. A valley called Ithaca

Fig. 10.8 Mimas, one of Saturn's moons about 400 km in diameter. The large crater at the *top* is Herschel crater, with walls about 10 km high (Credit: NASA).

Chasma stretches three quarters the way around the moon and in places is up to 100 km wide. The Chasma may have been formed by the Odysseus impact or by tectonic activity.

Dione is 377,400 km from Saturn with a diameter of 1120 km. It too is heavily cratered with a number of large impact craters scattered over its surface. There are also some smooth areas that may be due to coverings of water ice. Recent images show the surface contains features such as faults, valleys, and depressions caused by tectonic movement. The largest craters are about 100 km across, and bright streaks that are seen radiating from some craters are the result of material being ejected after impact.

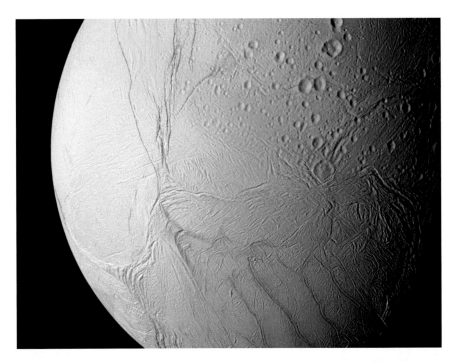

Fig. 10.9 Surface of the moon Enceladus taken by Cassini. The surface is mostly cratered and crossed by numerous fault lines. The Cassini spacecraft has found evidence of icy water particles spewing from geysers on Enceladus (Credit: NASA).

Rhea is Saturn's second largest moon, being 1530 km in diameter. This moon is 527,040 km from Saturn. The surface is very old and shows little change by geological activity. Like Dione, parts of its surface are heavily cratered and parts are smooth. There are also some white streaks across the surface that may be ice-filled cracks. Rhea also has two very large impact basins on its anti-Saturnian hemisphere, which are about 400 and 500 km across. Rhea's density of 1.23 times that of liquid water suggests that Rhea is three quarters ice and one-quarter rock (with some ammonia). In November 2010, NASA announced the discovery of a tenuous atmosphere on Rhea. It consists of oxygen and carbon dioxide in proportion of roughly 5–2. The main source of oxygen is thought to be radiolysis of water ice at the surface by ions supplied by the magnetosphere of Saturn. The source of the carbon dioxide is less clear, but it may be related to oxidation of the organics present in ice or to outgassing of the moon's interior.

Fig. 10.10 Close up view of the cratered surface of the moon Iapetus as seen by Cassini. The large impact crater near the bottom is Engelier (Credit: NASA/ESA).

Iapetus is the third largest moon orbiting Saturn but it is one of the most unusual. It has a diameter of 1440 km and orbits Saturn at a distance of 3,561,300 km from Saturn. One side of this moon is cratered and bright (probably due to water ice), while the other side is coated with a dark (probably carbon-based) material whose origin is not known. The dividing line between the two regions is relatively sharp. The Cassini probe revealed Iapetus to be heavily cratered with a bizarre patchwork of pitch-dark and snowy-white regions. The pole regions of Iapetus are as bright as its trailing hemisphere. Cassini discovered a 20 km tall equatorial ridge, which spans nearly the moon's entire equator (see Fig. 10.10).

Most of the remaining moons of Saturn range in diameter from 3 km to 255 km. Many of these moons are irregular in shape and have unusual orbits, suggesting they are fragments of larger bodies or captured asteroids.

Hyperion is the largest of Saturn's minor moons; it has an extremely irregular shape, and a very odd, tan-colored icy surface resembling a sponge. The surface of Hyperion is covered with numerous impact craters, most with diameters only 2–10 km. It is the only moon known to have a chaotic rotation, which means Hyperion has no well-defined poles or equator.

The moon **Phoebe** orbits in a direction opposite to the orbits of the other moons and opposite to the direction of Saturn's rotation. Phoebe is roughly spherical and has a diameter of 213 km. Phoebe rotates on its axis every 9 h and completes a full orbit around Saturn in about 18 months. Its surface temperature is -198 °C. Phoebe also has a highly inclined orbit and its surface is covered with a dark material. Spectroscopic measurement showed that the surface is made of water ice, carbon dioxide, phyllosilicates, organics and possibly iron bearing minerals. Phoebe is believed to be a captured object that originated from the Kuiper belt. It also serves as a source of material for the largest known ring of Saturn (see Fig. 10.12).

Pan is the innermost of Saturn's known moons at a distance of 134,000 km. With a diameter of only 20 km, it was discovered on Voyager photographs in 1990. Pan orbits within the Encke Division (in Saturn's A ring). Small moons near the rings produce wave patterns in the rings.

Janus and **Epimetheus** are two irregularly shaped moons that are co-orbital (orbit together). These two bodies are separated by less than 100 km and their velocities are nearly equal. Their gravitational interaction causes them to exchange orbits every 4 years. Astronomers believe they are probably fragments of a single-body, now destroyed. Epimetheus and Janus are the fifth and sixth moons in distance from Saturn. Both are phase locked with their parent so that one side always faces toward Saturn. Being so close, they orbit in less than 17 h.

A faint dust ring is present around the region occupied by the orbits of Epimetheus and Janus, as revealed by images taken by the Cassini spacecraft in 2006. The ring has a radial extent of about

Fig. 10.11 The Cassini space probe took this photo of Epimetheus in December 2007. The moon is 116 km in diameter and irregular in shape. Heavy cratering on its surface indicates it may be several billion years old (Credit: NASA).

5000 km. Its source is thought to be particles blasted off their surfaces by meteoroid impacts, which have formed a diffuse ring around their orbital paths.

All of Saturn's regular moons except Iapetus orbit very nearly in the plane of Saturn's equator. The outer irregular satellites follow moderately to highly eccentric orbits, and none is expected to rotate synchronously as all the inner moons of Saturn do (except for Hyperion). The exact number of Saturnian moons cannot be given, because there is no distinction between the countless small anonymous objects that form Saturn's ring system and the larger objects that have been named as moons. Over 150 moonlets embedded in the rings have been detected by the disturbance they create in the surrounding ring material, though this is thought to be only a small sample of the total population of such objects.

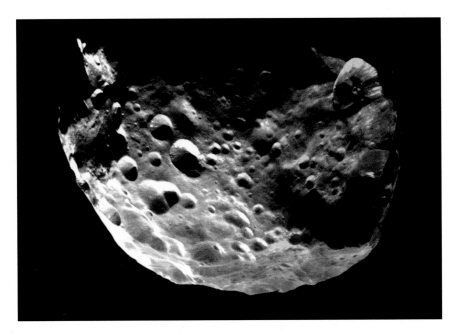

Fig. 10.12 The moon Phoebe has a diameter of 220 km but orbits in a direction opposite to that of the other moons (Credit: NASA).

Chariklo

Chariklo is an asteroid that orbits the Sun between the orbits of Saturn and Uranus at an average distance from the Sun of 15.87 AU. James Scotti of the Spacewatch program discovered the object on 15 February 1997. It takes 63.17 years to orbit the Sun and has a diameter of 250 km. A stellar occultation in 2013 revealed that Chariklo has two rings, one about 7 km wide and the other about 3 km wide, at 396 km and 405 km from Chariklo. This makes it the smallest known object to have rings. The existence of a ring system around an asteroid was unexpected because it had been thought that rings could only be stable around much more massive bodies. The origin of these rings remains a mystery, but they may be the result of a collision that created a disk of debris.

Further Information

http://nssdc.gsfc.nasa.gov/planetary/planetfact.html
www.space.com/saturn/
https://solarsystem.nasa.gov/planets/profile.cfm (check on Saturn)

11. Uranus: The Coldest Planet

Highlights

- Uranus is the coldest planet with a faint ring system and an extensive family of moons. The rings are composed of particles ranging in size from fine dust to several metres in diameter.
- Uranus has an odd magnetic field that is tilted at 59° to its axis of rotation. The magnetic field does not even pass through the centre of the planet.
- In August 2014 scientists using the Keck telescope in Hawaii photographed several huge bright storms on Uranus.
- Most of the moons of Uranus are quite dark possibly due to radiation darkening of methane on their surfaces.

Uranus is the third of the gas giants and the seventh planet from the Sun. The planet is the third largest in the solar system with a diameter of 51,118 km. Uranus is one-third the diameter of Jupiter but four times the diameter of Earth. It is large enough to hold about 64 Earths. Its distance from the Sun varies from about 3010 million km at its farthest point to about 2739 million km at its closest point. Because of this large distance, it takes Uranus 84 years to orbit the Sun once. Since its discovery in March 1781 it has only gone around the Sun just over two and half times. Light from the Sun takes just over 8 min to reach Earth, but it takes about 2 h 40 min to reach Uranus.

Uranus travels around the Sun in a slightly elliptical orbit and it spins on its axis once every 17 h 14 min. It is the furthest planet that can be seen with the unaided eye from Earth, however, it is faint and difficult to detect. Uranus was the first planet to be discovered by telescope.

Uranus is pale blue in colour with few surface features. It is the coldest planet and has a faint ring system and an extensive

J. Wilkinson, *The Solar System in Close-Up*, Astronomers' Universe,
DOI 10.1007/978-3-319-27629-8_11,
© Springer International Publishing Switzerland 2016

Fig. 11.1 The planet Uranus as seen by the Hubble Space telescope (Credit: NASA).

family of moons. Little was known about the planet until the Voyager 2 probe flew by it in January 1986 (Fig. 11.1).

It is possible to see Uranus without a telescope from Earth. At magnitude 5.3, Uranus is just within the brightness scale that a human eye can perceive. Of course, you've got to have extremely dark skies and know exactly where to look to see Uranus.

Early Views About Uranus

Uranus is the ancient Greek deity of the Heavens, the earliest supreme god. Uranus was the son and mate of Gaia the father of Cronus (Saturn) and of the Cyclopes and Titans (predecessors of the Olympian gods). Uranus was not seen in the night sky by ancient observers because it is so far away from the Sun and therefore faint. The first recorded sighting of this object was made in 1690 by English astronomer, Joh Flamsteed, who

Table 11.1 Details of Uranus

Distance from Sun	2,870,990,000 km (19.2 AU)
Diameter	51,118 km
Mass	8.68×10^{25} kg (14.53 Earth's mass)
Density	1.29 g/cm^3 or 1290 kg/m^3
Orbital eccentricity	0.046
Period of revolution	30,685 Earth days or 84.01 Earth years
Rotation period	17 h 14 min
Orbital velocity	24,516 km/h
Tilt of axis	97.86°
Average temperature	−200 °C
Number of Moons	At least 27
Atmosphere	Hydrogen, helium, methane
Strength of gravity	8.2 N/kg at surface

incorrectly cataloged it as 34 Tauri. Another observer, Pierre Lemonnier, recorded Uranus as a star a total of 12 times during 1769.

William Herschel officially discovered Uranus in 1781 while observing the night sky through one of his telescopes. Herschel noted that the object moved against the background of stars over several nights and had a bluish-green disc, unlike the stars, which were point sources of light. Herschel concluded that the faint object was a planet and he called it "the Georgium Sidus" (the George's star) in honour of his patron, King George III of England; other people called the planet 'Herschel'. At that time Saturn was the furthest know planet in the solar system. Uranus was the first planet to be discovered with a telescope and its discovery effectively doubled the size of the known Solar System. Herschel also discovered the two larger moons of Uranus, Oberon and Titania.

Johann Bode, a German astronomer, named the planet Uranus after the Greco-Roman god who personified the universe and was the father of Saturn. The planet was officially named Uranus in 1850 (Table 11.1).

Probing Uranus

We know little about Uranus because it is so far from Earth. Most of what we do know came from the Voyager 2 probe, which flew by Uranus in early 1986. The probe passed within 82,000 km of the

Table 11.2 Significant space probes to Uranus

Probe	Country of origin	Launched	Notes
Voyager 2	USA	1977	Fly by in Jan 1986

planet's cloud tops. It took Voyager 2 five years to travel from Saturn to Uranus.

Photographs taken by Voyager 2 revealed Uranus was a blue coloured planet with a few faint bands of clouds moving parallel to its equator. There were no signs of belts or storm spots. Ten additional moons were discovered around the planet and most have at least one shattering impact crater. Voyager 2 also found that the planet's magnetic field was 50 times stronger than Earth's field. The magnetic field is tilted at 59° to its axis of rotation, and does not even pass through the centre of the planet. Because of this strange arrangement, the magnetic field wobbles considerably as the planet rotates (Table 11.2).

The possibility of sending the Cassini spacecraft to Uranus was evaluated during a mission extension planning phase in 2009. However, it would have taken about 20 years to get to the Uranian system after departing Saturn. A Uranus orbiter and probe was recommended by the 2013–2022 Planetary Science Decadal Survey published in 2011; the proposal envisages launch during 2020–2023 with a 13-year trip to Uranus. The ESA also looked at a "medium-class" mission called Uranus Pathfinder.

Position and Orbit

Uranus is the third largest planetary member of the solar system and is the seventh planet from the Sun. Its orbit is slightly elliptical and lies between Saturn and Neptune. Uranus has a mean distance from the Sun of about 2871 million km, placing it about 19.2 times farther from the Sun than Earth. It travels around the Sun once every 84.01 years and rotates on its axis with a period of 17 h 14 min. The planet's atmosphere rotates faster than its interior. The fastest winds on Uranus, measured about two-thirds of the way from the equator to the south pole, blow at about 720 km/h.

Most planets spin on their axis nearly perpendicular to its orbital plane, but Uranus' axis is almost parallel to this plane (nearly 98° to the vertical). This means the rotational axis is almost 8° below the orbital plane, so the planet appears to be tipped on its side. It is not known why Uranus has such a high axial tilt, but it may have been hit by another large body sometime in its past. At the time of Voyager 2's flyby, the planet's south pole was pointed almost directly at the Sun. As a result of this, Uranus' polar regions receive more energy from the Sun than its equatorial regions. However, the temperature is still hotter at the equator than at its poles, for unknown reasons.

Density and Composition

Uranus is a gaseous, icy planet with a mass about 14 times that of Earth, but only about one twentieth that of the largest planet Jupiter. The average density of Uranus is about 1.3 g/cm^3, which is about one quarter that of Earth. Thus the material Uranus is made out of must be light and icy.

In contrast to the other gas planets (Jupiter and Saturn), the composition of Uranus is not dominated by hydrogen and helium. Hydrogen accounts for only 15 % of the planets mass. Most of the planet is made up of methane, ammonia and water. There are three layers or regions inside the planet. The dense core (30 %) contains silicate/iron-nickel rock and various ices, but no liquid metallic hydrogen. The mantle (40 %) is probably highly compressed water ice with some methane and ammonia. The outer layer (30 %) lies at the base of the atmosphere, is considered to be ocean, and composed of mostly gaseous or liquid hydrogen, helium and methane (see Fig. 11.2).

The strength of gravity on Uranus is actually less than Earth's gravity (8.2 N/kg compared to Earth's 9.8 N/kg). This means that a 75 kg person, who weighs 735 N on Earth, would weigh only 615 N on Uranus.

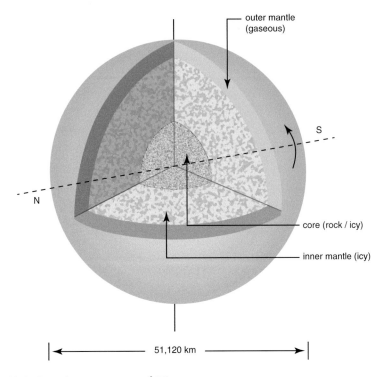

Fig. 11.2 Interior structure of Uranus.

The Surface

Being a gaseous planet, there is no solid surface layer on Uranus. The outer layer of the planet is made up of icy molecules of water, methane and ammonia. The surface we see from Earth is actually Uranus's atmosphere. Clouds are also visible in the atmosphere. The planet radiates about the same energy as it receives from the Sun and has little internal heat.

Uranus's internal heat appears markedly lower than that of the other giant planets. The lowest temperature recorded in Uranus's tropopause is −220 °C, making Uranus the coldest planet in the solar system. One of the hypotheses for this coldness suggests that a super massive impactor that caused it to expel most of its primordial heat hit Uranus. This impact left the planet with a depleted core temperature. Another hypothesis is that some form of barrier exists in Uranus's upper layers that prevent the core's heat from reaching the surface.

The Atmosphere

The atmosphere of Uranus is composed of mainly of hydrogen (83 %), and helium (15 %), and ices (such as water, ammonia, and methane), along with traces of hydrocarbons. The planet has the coldest atmosphere in the solar system, with a minimum temperature of −220 °C. The methane that is trapped high in the atmosphere absorbs red light from the visible spectrum, and this makes the planet appear blue-green in colour.

Along with methane, trace amounts of various hydrocarbons are found in the stratosphere of Uranus, which are thought to be produced from methane by photolysis induced by the solar ultraviolet radiation. The hydrocarbons include ethane, acetylene, methyl acetylene, and diacetylene. Spectroscopy has also uncovered traces of water vapor, carbon monoxide and carbon dioxide in the upper atmosphere, which can only originate from an external source such as infalling dust and comets.

Voyager 2 data showed the atmosphere contains three distinct cloud layers. The top layer contains ammonia, the next layer ammonium hydrosulfide, and the third or lower layer contains water ice. These layers are found deep in Uranus's atmosphere where temperature and pressures are higher. The atmospheric pressure beneath the cloud layer is about 1.3 times that at the Earth's surface (see Fig 11.3).

Like the other gas planets, Uranus has bands of clouds that blow around rapidly parallel to the equator. These bands are very faint and can only be seen with image enhancement of the Voyager 2 photographs. Winds at mid-latitudes are propelled in the rotational direction of the planet. Winds at equatorial latitudes blow in the opposite direction.

Recent observations with the Hubble Space Telescope show larger and more pronounced streaks and some spots. The spots are probably violent swirling storms like a hurricane.

Astronomers have also been able to chart the wind speeds on Uranus, and have found they can travel at 250 m/s.

In August 2014 scientists using the Keck telescope in Hawaii spotted huge storms on the planet Uranus. One image, taken in infrared light on 5 August, shows a few storms as bright spots in

Fig. 11.3 This infrared image taken by the Hubble ST allows astronomers to probe the structure of Uranus' atmosphere. The *red* around the planet's edge represents a very thin haze at a high altitude. The *yellow* is another hazy layer. The deepest layer, the *blue* on the *right*, shows a clearer atmosphere. The rings have been brightened in this image; in reality, the rings are as dark as charcoal. The bright spots are violent storms (Credit: HST/NASA).

photos taken of the planet. A second photo of Uranus, taken on 6 August, reveals many more bright spots. One very large storm seen by the telescope has particularly interested researchers analyzing the views because it reaches into the high altitudes of the planet's atmosphere. The swirling clouds and violent winds are being driven by massive bands of jet streams that can surround the entire planet. The planet's strange tilt also contributes to such bizarre weather systems (see Fig. 11.4).

Fig. 11.4 Scientists using the 10 m Keck Telescope in Hawaii spotted huge bright storms on Uranus in August 2014 (Credit: Imke de Pater (UC Berkeley)/Keck Observatory).

The Rings

Uranus has a number of thin rings surrounding it. These were discovered by chance in 1977 when Uranus appeared to pass in front of the faint star SAO158687 in the constellation of Libra, as seen from Earth. The rings temporarily interrupted light from the star and pulses of starlight were seen each side of the planet, suggesting there was something around the planet. In 1986, Voyager 2 confirmed the existence of a ring system. The ring system lies in the Uranian equatorial plane, circling Uranus between 38,000 km and 52,000 km from its centre.

The rings are faint and composed of particles ranging in size from fine dust to several metres in diameter. Voyager 2 found that the gaps between the rings were not empty but contained fine dust that may have originated from collisions between the larger particles forming the main rings or from the surrounding moons. The matter in the rings may once have been part of a moon (or moons) that was shattered by high-speed impacts.

The outer ring is the most massive, and its particles are kept in orbit by the gravitational influence of two moons, Cordelia and Ophelia. The outermost ring is about 100 km wide but only 10–100 m thick. The ring material is probably composed of chunks of ice, covered by a layer of carbon (see Fig. 11.5).

Fig. 11.5 Voyager 2 image of the rings of Uranus showing slight colour differences. The bright white ring at the bottom is the furtherest ring from Uranus—called Epsilon (Credit: NASA).

In December 2005, the Hubble Space Telescope detected a new pair of rings located twice as far from Uranus as the previously known rings. These new rings are so far from Uranus that they are called the "outer" ring system. Hubble also spotted two small satellites, one of which, Mab, shares its orbit with the outermost newly discovered ring. The new rings bring the total number of Uranian rings to 13. In April 2006, images of the new rings from the Keck Observatory showed the outermost is blue and the other one red. One hypothesis concerning the outer ring's blue color is that it is composed of minute particles of water ice from the surface of Mab that are small enough to scatter blue light. In contrast, Uranus's inner rings appear grey.

The Uranian rings were the first to be discovered after Saturn's. This was an important finding since we now know that rings are a common feature of large planets, not a peculiarity of Saturn alone. Uranus's rings are much darker than those of Saturn and harder to see from Earth. In 2014, astronomers reported that the rings of Uranus are probably very young, forming relatively recently, and not when the planet itself formed.

Temperature and Seasons

The temperature in the upper atmosphere of Uranus is about −200 °C. At this low temperature methane and water condense to form clouds of ice crystals. Because methane freezes at a lower temperature than water, it forms higher clouds over Uranus. Methane absorbs red light, giving Uranus its blue-green colour.

In the interior, the temperature rises rapidly to about 2300 °C in the mantle and about 7000 °C in the rocky core. The pressure in the core is about 20 million times that of the atmosphere at the Earth's surface.

The planet radiates as much heat as it receives from the Sun, back into space. Because its axis is tilted at 98°, its poles receive more sunlight during a Uranian year than does its equator. However, the weather system seems to distribute heat fairly evenly over the planet.

As the planet orbits the Sun, its north and south poles alternately point directly toward or directly away from the Sun, resulting in exaggerated seasons. During summer near the north pole, the Sun is almost directly overhead for many Earth years. At the same time southern latitudes are subjected to a continuous frigid winter night. Forty-two years later, the situation is reversed.

In August 2006, the Hubble Space Telescope captured images of a huge dark cloud on Uranus. The cloud measured about 1700 km by 3000 km. Scientists are not certain about the origin of the cloud.

Magnetic Field

Uranus has a magnetic field about 50 times stronger than Earth's. The axis of the field (an imaginary line joining its north and south poles) is tilted 59° from the planet's axis of rotation. The centre of the magnetic field does not coincide with the centre of the planet—it is offset by almost one third of Uranus's radius or nearly 7700 km. Because of the large angle between the magnetic field and its rotation axis, the magnetosphere of Uranus wobbles considerably as the planet rotates. One hypothesis is that, unlike the magnetic fields of the terrestrial and gas giants, which are generated within their cores, the ice giants' magnetic fields are generated by motion at relatively shallow depths, for instance, in the water–ammonia ocean. Another possible explanation for the magnetosphere's alignment is that there are oceans of liquid diamond in Uranus's interior that would deter the magnetic field.

Voyager 2 passed through Uranus's magnetosphere as it flew by Uranus. The magnetic field traps high energy, electrically charged particles, mostly electrons and protons, in radiation belts that circle the planet. As these particles travel back and forth between the magnetic poles, they emit radio waves. Voyager 2 detected these waves, but they are weak and cannot be detected from Earth. Uranus also has relatively well developed auroras, which are seen as bright arcs around both magnetic poles (Fig. 11.6).

The magneto tail of Uranus was measured by Voyager 2 to extend 10 million km out into space. Unlike other magneto tails, the magnetic field lines in the tail are cylindrical and appear

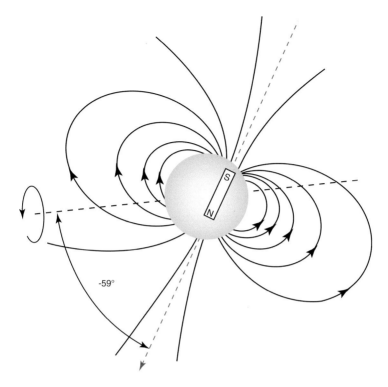

Fig. 11.6 Uranus' strange magnetic field.

wound around each other—similar to a corkscrew. This is probably due to the strange axial tilt of the planet.

Moons

Before the visit of Voyager 2 to Uranus in January 1986, only five moons were known to orbit Uranus. These moons were discovered between 1787 and 1948 and they range in size from 480 km to 1550 km in diameter. Voyager discovered ten more moons around Uranus, all less than 50 km in diameter. Several of these tiny, irregularly shaped moons are shepherd satellites whose gravitational pull confines them within the rings of Uranus. Astronomers used Earth-based telescopes to find two more moons in 1997 and three more in 1999.

Table 11.3 The eight largest moons of Uranus

Name of moon	Distance (km)	Period (days)	Diameter (km)	Discovered (year)
Portia	66,000	0.51	135	1986
Puck	86,000	0.76	162	1986
Miranda	130,000	1.41	471	1948
Ariel	191,000	2.52	1157	1851
Umbriel	266,000	4.14	1170	1851
Titania	436,000	8.71	1578	1787
Oberon	583,000	13.46	1522	1787
Sycorax	12,180,000	1288	190	1997

Unlike the other bodies in solar system that have names from classical mythology, the names of Uranus's moons are derived from the writings of Shakespeare and the English writer Alexander Pope.

Uranus has at least 27 moons, arranged in three groups: 13 small dark inner ones, five large moons, and nine more distant ones recently discovered by telescopes. Most of these moons have nearly circular orbits in the plane of Uranus's equator. The outer moons have elliptical orbits and many have retrograde motion. It may be that some of the smaller moons, especially the outer ones, are captured asteroid-like bodies (see Table 11.3).

Most of the moons of Uranus are quite dark. This may be due to radiation darkening of methane on their surfaces. The Uranian satellites have relatively low albedos; ranging from 0.20 for Umbriel to 0.35 for Ariel. They are ice–rock conglomerates composed of roughly 50 % ice and 50 % rock. The ice may include ammonia and carbon dioxide.

The five largest moons of Uranus (Miranda, Ariel, Umbriel, Titania, and Oberon) have higher densities than expected (1.4–1.7 g/cm^3). This suggests that they may contain more than 50 % rock or silicates, with smaller amounts of water ice than Saturn's similar-sized moons.

Miranda is one of the most unusual moons because it has a surface like no other in the solar system. The older areas are relatively smooth, cratered plains. But other areas look as if they have been clawed and gouged by something. Miranda possesses fault canyons 20 km deep, with terraced layers, and a chaotic variation in surface ages and features. Astronomers think Miranda's core originally consisted of dense rock while its outer layers were mostly ice; this structure has at some time in the past been dramatically changed. Surface variations suggest an asteroid or tectonic activity may have shattered the moon, breaking its surface into several pieces—the pieces have since been reassembled by gravity, leaving behind great scars. Miranda is just one-seventh the size of Earth's moon and is not quite spherical. It takes only 1.41 days to orbit the planet (see Fig. 11.7).

Ariel is a young moon with few craters, most being less than 50 km in diameter. It takes 2.52 days to orbit Uranus. The surface suggests there has been a lot of geological activity with many faults, fractures and valleys visible. Photographs taken by Voyager

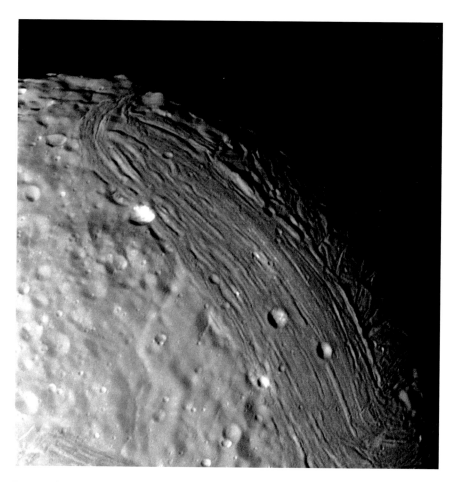

Fig. 11.7 A Voyager 2 photo of the surface of Miranda (Credit: NASA).

Fig. 11.8 A Voyager 2 photo of the surface of Ariel (Credit: NASA).

2, suggest many of the features are volcanic in origin and that some form of viscous fluid once flowed across the surface. Many of the fissures have been partially filled in with frozen deposits of an unknown material (see Fig. 11.8).

Umbriel is the darkest of the five larger moons orbiting Uranus. It takes 4.14 days to orbit Uranus. The surface of this moon is almost uniformly covered with craters. Many large craters suggest the surface is fairly old, although the covering of dark material appears to hide many features.

Titania is the largest of the Uranian moons, with a diameter of 1578 km. It takes 8.71 days to orbit Uranus. The surface of this moon contains numerous impact craters, with some young valleys, faults and fractures. One heavily fractured region is thought to have been caused by the crust fracturing as water froze below the surface (see Fig. 11.9).

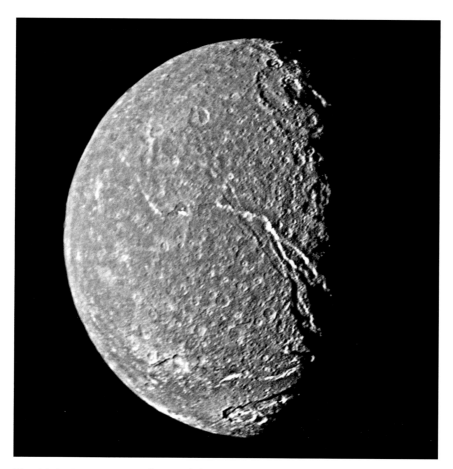

Fig. 11.9 A Voyager 2 photo of the surface of Titania (the largest moon of Uranus) (Credit: NASA).

Oberon has a diameter of 1522 km, making it slightly smaller than Titania. The moon takes 13.46 days to orbit Uranus. Its surface contains many impact craters some of which are surrounded by bright rays. One mountain towers above the surrounding terrain to a height of 20 km (see Fig. 11.10).

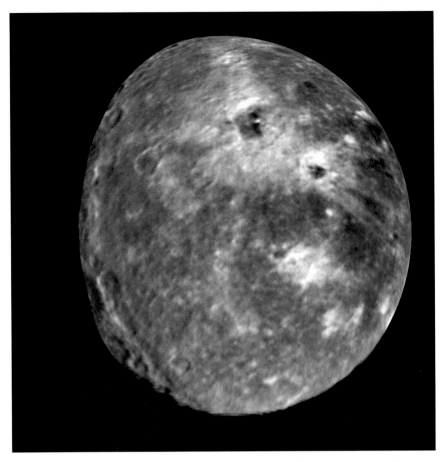

Fig. 11.10 The moon Oberon is Uranus's second largest moon (Credit: NASA).

Further Information

For fact sheets on any of the planets including Uranus check out
http://nssdc.gsfc.nasa.gov/planetary/planetfact.html
www.space.com/uranus/
https://solarsystem.nasa.gov/planets/ (click on Saturn)

12. Neptune: Another Cold World

Highlights

- The Voyager 2 probe found Neptune to be a large blue planet, with many markings, cloud bands and a system of faint rings.
- The planet Neptune has a profound impact on a region directly beyond it, known as the Kuiper belt.
- Neptune's atmosphere is very active with rapid changes in weather occurring regularly, and westward moving winds reaching speeds up to 2000 km/h (the fastest of all the planets).
- Neptune is a source of both continuous emission and irregular bursts of radio waves.
- Neptune has 14 moons; the most recent was discovered in 2013.

Neptune is the smallest of the gas giants and the eighth planet from the Sun. This planet is also the fourth largest in the solar system with a diameter of 49,532 km. Neptune is smaller in diameter than Uranus but it has more mass than Uranus. Although four times the size of Earth, Neptune has 17 times the mass of Earth but it is less dense than Earth.

Neptune is about 4.5 billion km from the Sun, making it about 1.5 times more distant than Uranus. It takes over 165 Earth years to orbit the Sun once, and it rotates on its axis once every 16.1 h. Because of its great distance from the Sun, Neptune receives only 0.1 % of the sunlight that Earth receives-so the planet's surface is very cold.

The planet cannot be seen with the unaided eye from Earth, but it can be seen through a good telescope. It then appears as a tiny featureless disc that is barely distinguishable from a star. Most of

J. Wilkinson, *The Solar System in Close-Up*, Astronomers' Universe,
DOI 10.1007/978-3-319-27629-8_12,
© Springer International Publishing Switzerland 2016

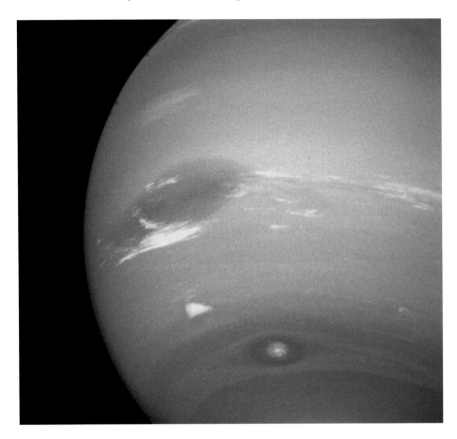

Fig. 12.1 Neptune as seen by Voyager 2. At the north (*top*) is the Great Dark Spot, accompanied by bright, white clouds that undergo rapid changes in appearance. To the south is the bright feature that Voyager scientists have nicknamed "Scooter." Still farther south is the feature called "Dark Spot 2," which has a bright core. Each feature moves eastward at a different velocity, so it is only occasionally that they appear close to each other, such as at the time this picture was taken (Credit: NASA).

our information about Neptune came from the Voyager 2 space probe that passed by Neptune in 1989. Voyager found Neptune to have a deep blue colour with an outer layer covered with whitish clouds (Fig. 12.1).

Early Views About Neptune

In Roman mythology Neptune was the god of the sea. Such a god was important to the Romans because sea travel formed a key part of their life. Neptune was the son of Saturn and Orps (called Cronus and Rhea by the Greeks). Neptune also resembled the Greek god Poseidon, which was god of the sea, earthquakes and horses.

Early astronomers did not know about Neptune as a planet because it could not be seen from Earth by the unaided eye. The discovery of Neptune was a great triumph of mathematics. After the discovery of Uranus in 1781, astronomers noticed that its orbit was not following the path predicted by Newton's and Kepler's laws of motion. Either these laws were wrong or there had to be another undiscovered planetary body in the solar system influencing Uranus' orbit. In 1845 and 1846, John Couch Adams in England and Urbain Le Verrier in France, independently calculated the mass and location of such a body. The body, now called Neptune, was discovered in 1846 by German astronomer Johann Gottfried Galle, close to the predicted position. Within a few weeks the British astronomer William Lassell discovered a moon around Neptune—it was called Triton after the half-man, half-fish son of the sea god (Table 12.1).

Table 12.1 Details of Neptune

Distance from Sun	4,504,000,000 km (30.1 AU)
Diameter	49,532 km
Mass	1.02×10^{26} kg (17.1 times Earth's mass)
Density	1.64 g/cm^3 or 1640 kg/m^3
Orbital eccentricity	0.010
Period of revolution	60,225 Earth days or 165 Earth years
Rotation period	16 h 7 min
Orbital velocity	19,548 km/h
Tilt of axis	29.6°
Average temperature	−225 °C
Number of Moons	14
Atmosphere	Hydrogen, helium, some methane
Strength of gravity	11.2 N/kg at surface

Probing Neptune

Astronomers know little about Neptune because it is so far from Earth. Neptune has been visited by only one spacecraft, Voyager 2 on 25 August 1989. Almost everything we know about Neptune has come from this encounter. Voyager had been travelling for about 12 years and had covered nearly 5 billion km to reach Neptune. The space probe came to within 5000 km of the planet and it collected a wealth of information about this most distant gas giant and its moons.

Voyager found Neptune to be a large blue planet, with many markings and cloud bands. Five thin rings were also found around the planet and six new moons were discovered to add to the two already known. The rings were found to be complete rings with bright clumps in them. One of the rings appeared to have a curious twisted structure. The rings are very dark and their composition is unknown (Table 12.2).

Position and Orbit

Neptune is the eighth planet from the Sun and the fourth largest planetary member of the solar system. Its orbit is slightly elliptical and lies beyond that of Uranus. Neptune has a mean distance from the Sun of about 4500 million km, placing it about 30.1 times further from the Sun than Earth. Neptune is so far away that when Voyager 2 was passing the planet the radio signals from the probe took over 4 h to reach Earth.

Neptune travels around the Sun once every 165 years and rotates with a period of 16 h 7 min. In its equatorial zone, winds blow westward at close to 1500 km/h, creating huge storms.

Neptune spins on its axis at an angle of 29° from the vertical. This amount of axial tilt is similar to that of Earth. Because of this

Table 12.2 Significant space probes to Neptune

Probe	Country of origin	Launched	Comments
Voyager 2	USA	1977	Passed by Neptune in August 1989

angle, it was expected that Neptune's poles would be colder than its equator. However, astronomers using Europe's Very Large Telescope in Chile, found that the planet's south pole is about 10 °C warmer than elsewhere. This imbalance in temperature probably explains why Neptune has such strong winds.

Neptune's orbit has a profound impact on the region directly beyond it, known as the **Kuiper belt**. The Kuiper belt is a ring of small icy worlds, similar to the asteroid belt but far larger, extending from Neptune's orbit at 30 AU out to about 55 AU from the Sun. Much in the same way that Jupiter's gravity dominates the asteroid belt, shaping its structure, so Neptune's gravity dominates the Kuiper belt. Over the age of the solar

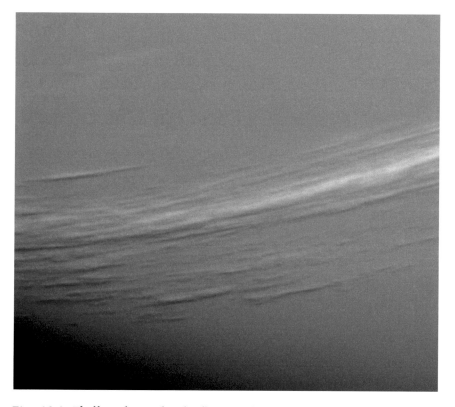

Fig. 12.2 Fluffy white clouds floating high in Neptune's atmosphere (Voyager 2) (Credit: NASA).

system, certain regions of the Kuiper belt became destabilised by Neptune's gravity, creating gaps in the Kuiper belt's structure.

Density and Composition

Neptune is a gaseous planet with a mass about 17 times that of Earth, but it is not as dense as Earth. The average density of Neptune is about 1.64 g/cm^3, compared to Earth's density of 5.52 g/cm^3. This is mainly due to the different composition of each planet—Earth is a rocky planet, while Neptune is a gaseous one.

Neptune's composition and interior is similar to that of Uranus. Both planets have a rocky core surrounded by frozen ammonia, methane and water. Hydrogen contributes only about 15 % of the planet's total mass. Compared to Jupiter and Saturn, Neptune has more ammonia, methane and water.

Neptune's interior consists of a small but dense core of melted rock, mostly iron, nickel and silicates. In the core, the temperature and pressure is very high. Although Neptune receives 40 % less sunlight than Uranus, their surface heat is almost same. More interestingly, Neptune gives off 2.6 times more energy than the energy it takes from the Sun! The rocky cores temperature is so hot that it can melt rocks. The high heat of Neptune and the cold temperature of the space around it create a huge temperature difference. This difference creates a huge wind blasting like a hurricane.

Because of Neptune's greater density, its core is probably slightly larger than the core of Uranus. Surrounding the core is a mantle of water, ammonia and methane, and an outer layer or ocean of hydrogen, helium and methane. However, Neptune's interior layers may not be distinct and uniform.

The strength of gravity on Neptune is greater than Earth's gravity (11.2 compared to Earth's 9.8 N/kg). This means that a 75 kg person weighing 735 N on Earth would weigh 840 N on Neptune.

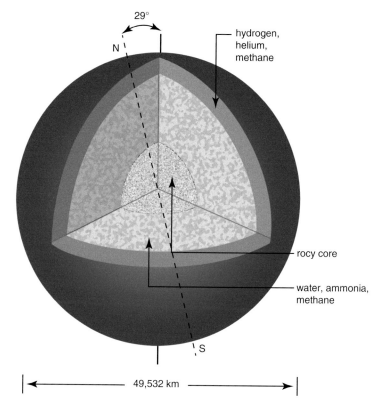

Fig. 12.3 Interior structure of Neptune.

The Surface

The surface of Neptune looks solid, but it is actually not solid. It is made out of gas. So what we actually see in pictures is not the surface of Neptune, but it is the top of the clouds of Neptune.

Most of the planet consists of compressed, frozen gases. The outer layer of the planet is best described as an ocean containing water mixed with methane and ammonia. Thick clouds cover the ocean so visibility would be difficult.

The Atmosphere

The atmosphere of Neptune contains a mixture of hydrogen, helium and some methane. Like the other gas giants, layers of ammonia, ammonium hydrosulfide, and water ice are also thought to exist. Methane in the top of the atmosphere gives the planet its blue colour. This colour is the result of absorption of red light by the methane.

Neptune's atmosphere is much more active than that of Uranus. Rapid changes in weather occur regularly, and westward moving winds reach speeds up to 2000 km/h (the fastest of all the planets). The winds are driven by the heat energy radiated out from Neptune's interior.

As Voyager 2 passed within 5000 km of the Neptunian cloud tops, its cameras revealed a wide variety of features. Bright polar collars and broad bands in different shades of blue were prominent in the southern hemisphere. Also visible were bright streaks of cirrus cloud stretched out parallel to the equator. The Voyager pictures also detected shadows of these clouds thrown onto the main cloud layer some 50 km below (see Fig. 12.2).

Voyager 2 also photographed large storms in the atmosphere. One in particular was about half the size of Jupiter's Great Red Spot, and oval in shape. Named the Great Dark Spot, it was observed to rotate anticlockwise over a period of about 10 days. Observation revealed this feature was a hole in the Neptunian clouds through which the lower atmosphere could be seen. Winds blew the spot westward at about 300 m/s. Cirrus-type clouds of frozen methane were seen forming and changing shape above and around the Great Dark Spot (see Fig. 12.4).

Voyager 2 also identified a smaller dark spot in the southern hemisphere and a small irregular white cloud that moves around Neptune every 16 h or so, known as the 'Scooter'. This latter feature may be a gas plume rising from lower in the atmosphere.

Various white streaks and spots have been detected on Neptune but most have disappeared or changed greatly, while other features have emerged. In 1994 images from the Hubble Space Telescope showed the Great Dark Spot had disappeared (in contrast to Jupiter's Great Red Spot, which has lasted hundreds of years). This indicates that Neptune's atmosphere changes rapidly.

Neptune is also covered with a number of belts and zones that are fainter than those on Jupiter. A broad, darkish band is prominent at high southern latitudes. Embedded in this band is a smaller dark spot, about the size of Earth. White and wispy clouds seem to hover over the smaller dark spot. Scientists believe the darker clouds on Neptune contain hydrogen sulfide.

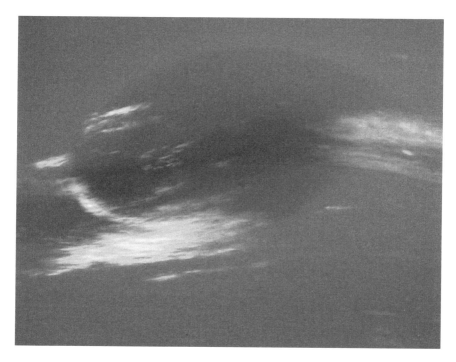

Fig. 12.4 Neptune's Great Dark Spot has a diameter about the size of Earth (Credit: NASA).

Neptune's spectra suggest that its lower atmosphere is hazy due to the condensation of products of ultraviolet photolysis of methane, such as ethane and acetylene. There are also trace amounts of carbon monoxide and hydrogen cyanide. Elevated concentrations of hydrocarbons make this layer warmer than expected.

For reasons that remain unknown, the planet's outer atmosphere layer has a high temperature of about 750 °C. The planet is too far from the Sun for this heat to be generated by ultraviolet radiation. One candidate for a heating mechanism is atmospheric interaction with ions in the planet's magnetic field. Other candidates are gravity waves from the interior that dissipate in the atmosphere.

The Rings

Astronomers on Earth thought Neptune might have some incomplete rings or arcs when they observed an occultation of a star by Neptune in 1984. However, pictures taken by Voyager 2 revealed

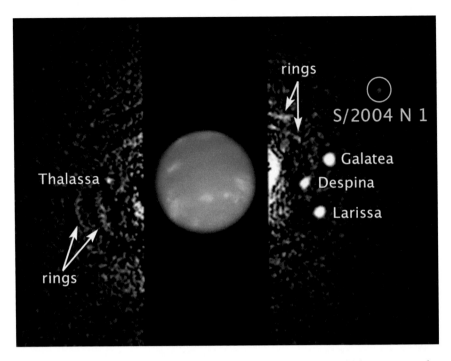

Fig. 12.5 Neptune's rings and moons as seen by the Hubble Space Telescope. The image of the planet was blocked out to capture detail in the rings, and then reinserted (Credit: NASA/ESA).

that these arcs were part of a narrow ring system that contains three areas. The three main rings are the narrow Adams Ring (63,000 km from the centre of Neptune), the Le Verrier Ring (at 53,000 km), and the broader but fainter Galle Ring (at 42,000 km).

The rings may consist of ice particles coated with silicates or carbon-based material, which most likely gives them a reddish hue. They are hard to see because they consist of small particles that reflect little light. Their exact composition is unknown but, because the temperature is so low, they probably contain frozen methane. Some of this methane has been changed by radiation into other carbon compounds, thus making the rings appear dark (see Fig. 12.5).

Temperature and Seasons

The large distance between Neptune and the Sun, means that the average temperature on Neptune is a very cold −225 °C. This temperature is low enough to freeze methane. Bright clouds of methane ice form in the upper atmosphere, and cast shadows on the lower cloud layers.

Since Neptune takes 165 years to orbit the Sun, the time between any seasons is very long (40 years) and temperatures do not vary much season to season. The planet does have a north and south pole and it's axis is tilted by nearly 30° from the vertical.

Because of seasonal changes, the cloud bands in the southern hemisphere of Neptune have been observed to increase in size and albedo. This trend was first seen in 1980 and is expected to last until about 2020.

Neptune's interior is believed to be very similar in composition and structure to Uranus. Core temperature is therefore expected to be around 7000 °C and core pressure about 20,000 atm (similar to Uranus).

Magnetic Field

Neptune has a strong magnetic field—about 25 times greater than Earth's. However, the magnetic axis is tilted 47° from Neptune's rotational axis and it is off-centre by more than half the radius of the planet. The magnetic field is probably generated in middle regions where the pressure is high enough for water to conduct electrical currents.

Neptune's bow shock, where the magnetosphere begins to slow the solar wind, occurs at a distance of 35 times the radius of the planet. The magnetopause, where the pressure of the magnetosphere counterbalances the solar wind, lies at a distance of 23–27 times the radius of Neptune. The tail of the magnetosphere extends out to at least 72 times the radius of Neptune, and likely much farther.

Observation of Neptune in the radio-frequency band shows that the planet is a source of both continuous emission and irregular bursts. Both sources are believed to originate from the planet's rotating magnetic field. In the infrared part of the spectrum, Neptune's storms appear bright against the cooler background, allowing the size and shape of these features to be readily tracked.

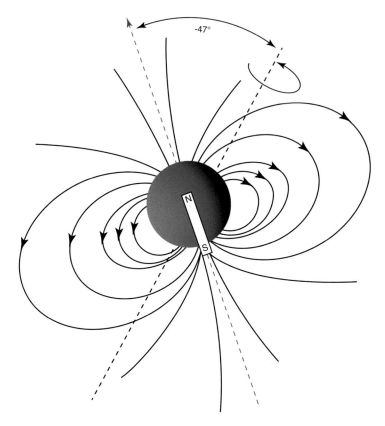

Fig. 12.6 Neptune's magnetic field.

Moons of Neptune

Prior to the Voyager 2 encounter, only two moons were known to exist around Neptune. Today we know Neptune has 14 moons, following the discovery of new moons by Voyager 2 and the Hubble Space Telescope. William Lassell discovered the largest moon, Triton, in 1846. Triton is spherical in shape and has a nearly circular orbit. The other moons are small, irregular shaped objects with highly elliptical orbits, suggesting Neptune has captured them. One of the interesting things about Triton is that it has a retrograde orbit; that is, it orbits in the opposite direction to Neptune's rotation. Some astronomers believe Neptune may also have captured Triton some 3 to 4 billion years ago.

Triton has a diameter of 2700 km and orbits Neptune every 5.88 days at a distance of 354,800 km. Its surface contains many interesting features including fault lines, cracks and ice (water, methane and ammonia) flows. There are not many impact craters, an indication that ice flows from the interior may have caused extensive resurfacing. The equatorial region contains a wrinkled terrain that resembles the skin of a cantaloupe or rock melon. Long narrow valleys rimmed by ridges cross the area. Such a region may have formed from repeated episodes of melting and cooling of the icy crust. Triton also has a few frozen lakes that may be the calderas of extinct ice volcanoes. In other areas, dark features surrounded by bright aureoles or rings are visible—these may be may be some of the geyser-like plumes detected by Voyager 2. The pinkish South Polar Region is covered by a cap of frozen methane and nitrogen and temperatures are around −245 °C. The pinkish colour of the polar cap is probably due to frozen nitrogen.

The very thin atmosphere of Triton contains nitrogen and methane.

Triton's density was found to be a little over 2 g/cm^3, roughly twice the density of water. This suggests that Triton is made up of a mixture of rock and icy material.

Triton has a tidal effect on Neptune that is tending to pull Triton towards Neptune. In about a quarter of a billion years, Triton may be pulled apart by Neptune's gravitational pull.

The inner four moons, Naiad, Thalassa, Despina and Galatea orbit within the ring system (see Fig. 12.7). Larissa, S/2004N1, Proteus, Triton, Nereid and the five small, outer moons orbit beyond the ring system. Four of Neptune's moons have retrograde motions—Triton, Halimede, Psamathe, and Neso. Many of the names of Neptune's moons come from the Nereids or water spirits of Greek mythology (Fig. 12.8).

Nereid's orbit is the most eccentric in the solar system. Its distance to Neptune ranges from about 1,353,600–9,623,700 km. The unusual and inclined orbit of Nereid suggests that it may be either a captured asteroid or Kuiper belt object, or that it was an inner moon in the past and was perturbed during the capture of Neptune's largest moon Triton. Prior to the visit of Voyager, Nereid was thought to be the second largest moon of Neptune. However, when Voyager discovered Proteus, Proteus was found to

Table 12.3 Moons of Neptune

Name of moon	Distance (km)	Period (days)	Diameter (km)	Discovered (year)
Naiad	48,200	0.29	66	1989
Thalassa	50,100	0.31	82	1989
Despina	52,500	0.34	150	1989
Galatea	62,000	0.43	176	1989
Larissa	73,500	0.56	194	1989
S/2004N1	105,283	0.936	18	2013
Proteus	118,000	1.12	420	1989
Triton	354,800	5.88	2700	1846
Nereid	5,510,000	360	340	1949
Halimede	15,728,000	1880	48	2002
Soa	22,422,000	2914	48	2002
Laomedeia	23,571,000	3167	48	2002
Psamathe	46,695,000	9115	28	2003
Neso	48,387,000	9373	60	2002

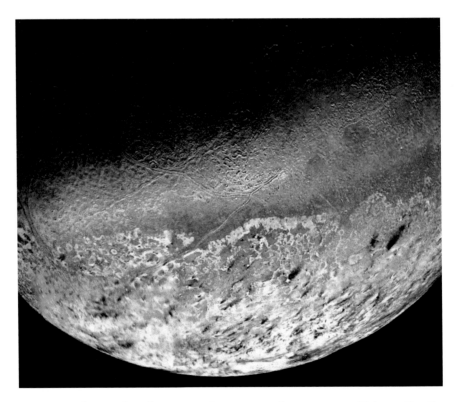

Fig. 12.7 The south pole region of Neptune's largest moon Triton. Credit: NASA.

Fig. 12.8 Proteus is a dark, cratered moon of Neptune (Credit: NASA).

be larger than Nereid. Apart from Triton, all the moons of Neptune are irregular in shape, all being too small for them to form uniform spheres (they are more asteroid-like).

In July 2013 NASA's Mark Showalter found a new moon orbiting Neptune by analysing archived photographs the Hubble Space Telescope captured between 2004 and 2009. The moon, designated S/2004N1, is estimated to be no more than 18 km across, making it the smallest known moon in the Neptunian system. It is so small and dim that it is roughly 100 million times fainter than the faintest star that can be seen with the naked eye. It even escaped detection by NASA's Voyager 2 spacecraft, which flew past Neptune in 1989. The designation 'S/2004N1' is provisional; '2004' refers to the year the data was first acquired, not the year of discovery.

Neptune's Status in the Solar System

From its discovery in 1846 until the subsequent discovery of Pluto in 1930, Neptune was the farthest known planet from the Sun. Upon Pluto's discovery Neptune became the penultimate planet, save for a 20-year period between 1979 and 1999 when Pluto's elliptical orbit brought it closer to the Sun than Neptune. The discovery of the Kuiper belt in 1992 led many astronomers to debate whether Pluto should be considered a planet in its own right or part of the belt's larger structure. When the International Astronomical Union defined the term 'planet' for the first time in 2006, Pluto was reclassified as a 'dwarf planet' and Neptune once again became the outermost planet in the solar system.

Further Information

http://nssdc.gsfc.nasa.gov/planetary/planetfact.html
www.space.com/neptune/
https://solarsystem.nasa.gov/planets/profile.cfm (check out Neptune)

13. Beyond Neptune: TNO's and Comets

Highlights

- Pluto is classified as a dwarf planet with at least five moons.
- The New Horizons probe flew within 10,000 km of Pluto in July 2015 and sent back some amazing up close images of the body.
- Kuiper belt objects that are also dwarf planets include: Pluto, Haumea and Makemake.
- Eris is a dwarf planet with one moon located in an outer part of the solar system known as the Scattered disc.
- Sedna has the longest orbital period of any known large object in the solar system, calculated at around 11,400 years. It exists in a region known as the Oort cloud.
- In 2014, the Rosetta spacecraft became the first spacecraft to orbit a comet and place a lander on its surface—spectacular pictures were returned.

Our understanding of the solar system has changed in the last decade mainly due to a new definition of what constitutes a planet by the International Astronomical Union, and a host of newly discovered objects that exist beyond the planet Neptune. Any object in the solar system that orbits the Sun at a greater distance on average than the planet Neptune has been termed a Trans-Neptunian Object (TNO). By 1 January 2008, astronomers had catalogued over a thousand Trans-Neptunian Objects and more have been detected since this date.

The first astronomer to suggest the existence of a Trans-Neptunian population was Frederick C. Leonard in 1930. In 1943, Kenneth Edgeworth postulated that in a region beyond Neptune, the material from the primordinal solar nebula would have been

J. Wilkinson, *The Solar System in Close-Up*, Astronomers' Universe,
DOI 10.1007/978-3-319-27629-8_13,

too widely spaced to condense into planets. From this he concluded that the outer region of the solar system should contain a very large number of smaller bodies that could from time to time, venture into the inner solar system. In the last few decades, astronomers have identified three regions that exist beyond Neptune: the Kuiper belt, the Scattered disc, and Oort cloud.

Trans-Neptunian Objects display a wide range of colours from blue-grey to very red. It is difficult to determine the size of TNOs because they are so far away. For large objects, diameters can be precisely measured during an occulation of a star. For smaller objects, diameters need to be estimated by thermal and relative brightness measurements. Orbital characteristics are also difficult to determine because these objects travel so slow.

The Kuiper Belt

The Kuiper belt is a vast region of the solar system beyond Neptune's orbit (see Fig. 13.1). It is best described as a flat dough-nut shaped disc that extends from 30 AU to 50 AU around the Sun. It is similar to the Asteroid belt, although it is much larger and more massive. Like the Asteroid belt, it contains mainly small bodies. But while the asteroids are composed mainly of rock and metal, the objects in the Kuiper belt are made mostly of rock and ices. The temperature of objects within the Kuiper belt is around $-230\ °C$, so they are very cold.

The Kuiper belt is named after Dutch-born American astronomer Gerard Kuiper, who first predicted its existence in 1951. The objects in this region are called Kuiper belt objects or KBOs. The orbits of many of these objects are highly elliptical and destabilised by Neptune's gravity.

The classical Kuiper belt appears to be made up of two separate populations. The first, known as the "dynamically cold" population, has nearly circular orbits, with relatively low inclinations to the ecliptic (up to about 10°). The second, the "dynamically hot" population, has orbits much more inclined to the ecliptic (by up to 30°). The two populations not only possess different orbits, but different colours; the cold population is markedly redder than the hot. If this is a reflection of different compositions, it suggests they

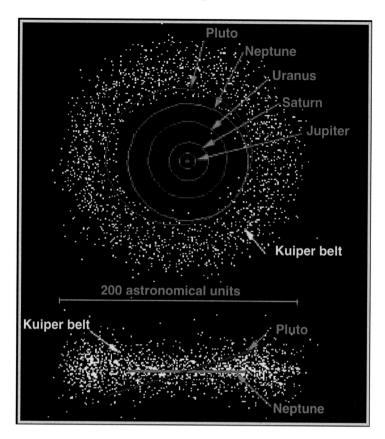

Fig. 13.1 The Kuiper belt is a region beyond the orbit of Neptune. The diagram shows the belt as seen from above the solar system and from side on (Credit: NASA).

formed in different regions. The hot population is believed to have formed near Jupiter, and to have been ejected out by movements among the gas giants. The cold population, on the other hand, has been proposed to have formed more or less in its current position, although it might also have been later swept outwards by Neptune. It has also been suggested the colour difference may reflect differences in surface evolution.

The most well known KBO is Pluto. Pluto was once regarded as a planet, but was reclassified in 2006 as a dwarf planet. Apart from Pluto, the first KBO was discovered by David Jewitt and Jane Luu in Hawaii in August 1992 and is called 1992 QB1. This object was found 42 AU from the Sun. Six months later, these two astronomers discovered a second object, 1993 FW, in the same region.

It is suspected that there may be as many as 35,000 objects in the Kuiper belt with diameters of 100 km or greater and many more smaller objects. Most objects in this belt probably formed at the same distance from the Sun as we find them today.

The Kuiper Belt is also thought to be the home of short period comets (those with periods less than 200 years). Comets are small bodies of ice and rock in orbit around the Sun. Many comets have highly elliptical orbits that occasionally bring them close to the Sun. When this happens the Sun's radiation vaporises some of comet's icy material, and a long tail is seen extending from the comet's head and pointing away from the Sun.

The former planet Pluto and its companion Charon are two of the larger KBOs. Several other large KBOs have been discovered, including Quaoar, Makemake and Orcus.

Pluto

Pluto was once classified as the ninth major planet of the solar system, but was reclassified in 2006 by the IAU as a dwarf planet. The main reason why Pluto was demoted as a major planet was that it has not cleared the neighbourhood around its orbit. Pluto orbits the Sun in the inner Kuiper belt where many other objects also orbit.

Pluto is much smaller than any of the official planets and is even smaller than seven of the moons in the solar system. It is so small and distant that we cannot see any surface detail on the planet through Earth based telescopes. It has a diameter of 2370 km (as measured by the New Horizons probe) and takes 248 years to travel once around the Sun. The other strange thing about this body is that its orbit is on a different plane to those of the major planets.

Pluto orbits the Sun at an average distance of about 5913 million km. It is so distant that Pluto's brightest daylight is less than moonlight on Earth. Pluto is always further from the Sun than Uranus, but every 248 years it moves inside Neptune's orbit for about a 20 year period, during which time it is closer to the Sun

Fig. 13.2 Image of Pluto taken by the New Horizons probe as it flew by Pluto in July 2015. *Left* of the bright *heart shaped* region are some impact craters (Credit: NASA/APL/SwRI).

than Neptune. Pluto crossed Neptune's orbit on 23 January 1979, and remained within it until 11 February 1999.

At its closest approach to the Sun, Pluto is 30 times more distant from the Sun than Earth. At its farthest distance from the Sun, Pluto is 50 times more distant from the Sun than Earth. Pluto will next be at its maximum distance from the Sun during the year 2113. During the coldest 124 years of its orbit, all of Pluto's atmosphere condenses and falls to the surface as frost. Images taken of Pluto by the Hubble Space Telescope have shown the reflectivity of its surface varies. Lighter areas are probably patches of nitrogen and methane frost as well as exposed regions of water ice.

Early Views About Pluto

In Roman mythology Pluto (Greek: Hades) was the god of the underworld. It also received this name because it was so far from the Sun and was in perpetual darkness.

Early astronomers did not know about Pluto because it could not be seen from Earth by the unaided eye. It is even difficult to locate using Earth-based telescopes.

The discovery of Pluto is an interesting story. Irregularities in the orbits of Uranus and Neptune led to the suggestion by US astronomers Percival Lowell and William Pickering that there might be another body (planet X) orbiting beyond Neptune. Lowell died in 1916, but he initiated the construction of a special wide field camera to search for planet X. In 1930, Clyde W. Tombaugh at Lowell Observatory in Arizona found planet X, which was later named Pluto. As it turned out, Pluto was too small and too distant to influence the orbits of Uranus and Neptune, and the search for another planet continued. The name 'Pluto' also honours Percival Lowell, whose initials PL are the first two letters of the name.

At one time it was thought that Pluto may have once been a moon of Neptune, but this now seems unlikely (Table 13.1).

Table 13.1 Details of Pluto

Distance from Sun	5,745,000,000 km (39.6 AU)
Diameter	2370 km
Mass	1.3×10^{22} kg (0.002 Earth's mass)
Density	1.8 g/cm^3 or 1800 kg/m^3
Orbital eccentricity	0.248
Period of revolution	90,740 Earth days or 249 Earth years
Rotation period	6.38 Earth days
Orbital velocity	17,064 km/h
Tilt of axis	122.5°
Average temperature	−220 °C
Number of Moons	5
Atmosphere	Nitrogen, methane
Strength of gravity	0.67 N/kg at surface

Probing Pluto

A spacecraft called New Horizons was launched on the 19 January 2006 on a mission to Pluto and the Kuiper belt. Using a combination of monopropellant and gravity assist, it flew by the orbit of Mars on 7 April 2006, Jupiter on 28 February 2007, the orbit of Saturn on 8 June 2008, and the orbit of Uranus on 18 March 2011. New Horizons flew within 10,000 km of Pluto in mid July 2015. The probe had a relative velocity of nearly 50,000 km/h at closest approach, and came as close as 27,000 km to Charon. After passing by Pluto, New Horizons continued farther into the Kuiper belt. Mission planners are hoping to flyby one or more additional Kuiper belt objects.

Position and Orbit

Pluto orbits the Sun in an elliptical path at an average distance from the Sun of about 5913 million km. Because it is so far from the Sun, Pluto takes a very long 248 years to go around the Sun once. The strange thing about Pluto is that its orbit is inclined at an angle of 17.2° to the orbital plane of the other planets. This means that its orbit rises and drops below the ecliptic plane. Pluto's plane is so elliptical that for 20 years of its orbital period it is closer to the Sun than Neptune (which follows a near circular orbit).

Like Uranus, Pluto is also tipped over on its side. Its rotational axis is inclined at an angle of 122.5° to the plane of its orbit. This means that Pluto's equator is almost at right angles to the plane of its orbit. Pluto also rotates in the opposite direction from most of the other planets, with one rotation taking 6 days 9 h and 18 min (Fig. 13.3).

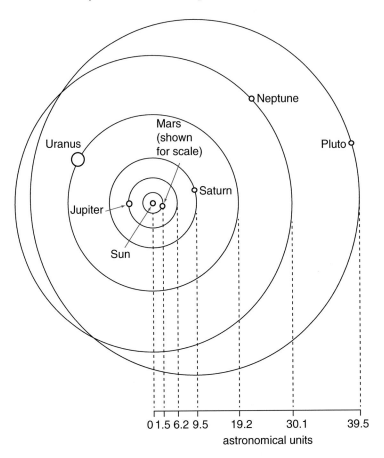

Fig. 13.3 Orbital path of Pluto.

Density and Composition

The average density of Pluto (just below 1.8 g/cm^3) indicates it's composition is a mixture of about 70 % rock and 30 % water ice, much like Triton. One theory is that Pluto and Triton formed at the same time in the same part of the solar nebula.

Pluto probably has a large rocky core of silicate materials, surrounded by a mantle rich in ices and frozen water. The extent of Pluto's crust is unknown, but it is thought to be covered with patches of frozen nitrogen, water, methane and ethane.

The strength of gravity on Pluto is much less than Earth's gravity (0.67 compared to 9.8 N/kg). This means that a 75 kg person weighing 735 N on Earth, would weigh only 50 N on Pluto.

The Surface

Astronomers know little about the surface of Pluto because the body is so far from Earth. Pluto's diameter is less than one fifth that of Earth's so it is difficult to observe anything on a surface that is so far away. The best views of Pluto show a brownish disc with bright and dark areas (see Fig. 13.2). The bright areas are probably covered with frozen nitrogen with smaller amounts of methane, ethane and carbon monoxide. The composition of the darker areas on the surface is unknown but they may be caused by decaying methane or carbon-rich material. Pluto's interior is probably rich in ices with some frozen water. The central core probably contains solid iron and nickel and rocky silicate materials. The New Horizons probe on approaching Pluto in 2015 noticed wide variations in the bright and dark areas on Pluto as well as a large, bright heart shaped feature. The probe found Pluto has a polar ice cap and ice mountains towering to 3500 m above plains of frozen methane and nitrogen. Infrared spectral images taken by New Horizons show dark patches representing concentrations of methane ice with striking differences in texture across different regions.

The Atmosphere

Little is known about Pluto's atmosphere but it thought to contain about 98 % nitrogen with about 2 % methane and carbon monoxide. The composition was determined from observations made when the planet passed in front of a bright star (an occulation).

The atmosphere is tenuous and the pressure at the surface is only a few millionths of that of Earth. It is thought to extend above the surface by about 600 km. Because of Pluto's elliptical orbit, the atmosphere may be gaseous when Pluto is near the Sun and frozen on the surface when furthest from the Sun. NASA wanted its New Horizon spacecraft to arrive at Pluto when the atmosphere was unfrozen. The probe found nitrogen escaping from Pluto's atmosphere.

Pluto's weak gravity means its atmosphere extends to a greater altitude than does Earth's atmosphere.

Temperature and Seasons

The surface temperature on Pluto varies between about −235 °C and −210 °C (very cold). The warmer regions roughly correspond to the darker regions on the surface. Pluto's surface appears darker when it is close to the Sun (as its atmosphere vaporises and exposes the dark surface), and brighter when furthest from the Sun (as its atmosphere condenses on the surface in a frozen state—which reflects light better).

Because the orbit of Pluto is so elliptical, the amount of solar radiation it receives varies markedly between its extreme positions. Its 248-year orbital period means that any seasonal change is very slow to take place.

Magnetic Field

Pluto may have a magnetic field, but it would not be strong. The New Horizons probe may be able to take measurements of any field that exists.

Moons of Pluto

In 1978 the American astronomer James Christy noticed that Pluto had an elongated shape in photographs. A search through previous images also showed a similar shape. This observation led to the discovery of a moon orbiting Pluto. The moon was named Charon and it was found to orbit Pluto at a distance of 19,700 km over a period of 6.39 days. It turned out that Pluto and Charon rotate synchronously, with Charon always facing Pluto. Astronomers were able to observe the two bodies rotating around each other during 1985 and 1990 when the two were edge-on to Earth. Such observations enabled astronomers to determine that Pluto's diameter is 2370 km and Charon's 1270 km. Because these sizes are close, some astronomers referred to the two bodies as a double-planet.

The average distance between Charon and Pluto is one-twentieth the distance between the Earth and our Moon. The combined mass of Pluto and Charon amount to less than one four-hundredth of Earth's mass.

The best pictures of Pluto and Charon have come from the New Horizons space probe that passed by these bodies in July 2015. Both objects are thought to consist of rock and ice. Charon's surface is probably covered with dirty water ice, which is why it doesn't reflect as much light as Pluto does. Photos from the New Horizon probe showed Charon has a dark north pole and a prominent impact crater near its equator that is ringed by bright rays. Deep canyons surround the crater, one larger than Earth's Grand Canyon. Charon's terrain appears to be younger than Pluto's surface (see Fig. 13.4).

In late 2005, a team of scientists using the Hubble Space telescope discovered two additional moons orbiting Pluto—they are now known as Nix and Hydra. These two tiny moons are

Fig. 13.4 Pluto's largest moon, Charon, as seen by the New Horizon's probe in July 2015. Charon has a swath of cliffs and troughs which stretch about 1000 km from *left* to *right*, which suggests widespread fracturing of its crust (Credit: NASA/JHUAPL/SwRI).

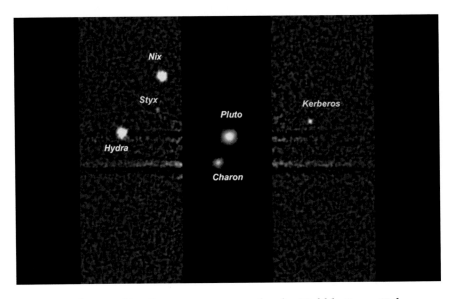

Fig. 13.5 Pluto and its five moons as seen by the Hubble Space Telescope. The *centre* of the image was blocked out (then reinserted) to allow for a longer exposure to capture the moons (Credit: NASA).

Table 13.2 Details of the moons of Pluto

Name	Distance from Pluto (km)	Period (days)	Diameter (km)	Discovered
Charon	19,571	6.39	1207	1978
Styx	42,393	20	20	2012
Nix	48,708	24.8	26	2005
Kerberos	57,729	32.1	28	2011
Hydra	64,698	38.2	30.5	2005

roughly 5000 times fainter than Pluto and between 48,000 km and 65,000 km away from Pluto.

A team led by Mark Showwalter from the SETI INSTITUTE in California discovered Pluto's smallest moons, Kerberos and Styx, in 2011 and 2012 respectively. Both were first seen in lengthy exposures of the Pluto system taken by the Hubble Space Telescope (see Fig. 13.5 and Table 13.2).

Other Kuiper Belt Objects

Since the 1980s, hundreds of icy bodies have been detected in the Kuiper belt. Most of these objects are much smaller than Pluto. Once the orbit of a KBO is determined it is given an official number by the Minor Planet Centre of the IAU. Once sufficient details of the body are determined, the object is given a name. Examples of KBOs with names include Asbolus, Bienor, Chaos, Chariklo, Chiron, Cyllarus, Deucalion, Elatus, Huya, Hylonome, Ixion, Nessus, Okyrhoe, Pelion, Pholus, Quaoar, Phadamanthus, Thereus, and Varuna. Full orbital details are known of only a few of the KBOs (see Table 13.3).

Ixion (2001 KX76) is a Kuiper Belt Object discovered on 22 May 2001. Its estimated diameter is 650 km and its distance from the Sun varies between 30 AU and 49 AU because of a highly elliptical orbit. It has a reddish colour and spectroscopic data suggests its composition is a mixture of water ice, dark carbon and tholin (a tar-like substance formed by irradiation of water and carbon-based compounds). Ixion and Pluto follow a similar but differently oriented orbit. Ixion's orbit is below the ecliptic, whereas Pluto's is above it. Ixion takes about 250 years to orbit the Sun.

Varuna (2000 WR106) is a small KBO named after the Hindu god of the sky, rain, oceans and rivers, law and the underworld. Varuna was discovered in November 2000 by R. McMillan and has a size of about 900 km. It orbits the Sun at an average distance of 43 AU in a near circular orbit. Unlike Pluto, which is in 2:3 orbital

Table 13.3 Main Kuiper Belt objects

Name	Diameter (km)	Orbital radius (AU)	Orbital period (years)	Orbital eccentricity	Number of moons
Pluto	2320	39.5	248	0.249	5
Ixion	650	30–49	250	0.242	0
Varuna	900	43	281	0.056	0
Quaoar	1110	43	286	0.039	1
2002AW197	750–768	41–53	325	0.132	0
Haumea	1240	43	283	0.195	2
Orcus	760–810	30–48	245	0.227	1
Makemake	1430	45.8	308	0.159	0

resonance with Neptune, Varuna is free from any significant perturbation from Neptune.

Varuna has a fast rotational period of about 6 h. The surface of this body is red but dark compared with other KBOs, suggesting the surface is largely devoid of ice.

One of the largest KBOs is **Quaoar** (2002 LM60) discovered in 2002 by astronomers Chad Trujillo and Mike Brown in California, USA, using large ground-based telescopes. The name Quaoar is derived from a group of Native American Tongva people, native to the area around Los Angeles, where the discovery was made. Quaoar is reddish in colour and orbits the Sun once every 286 years in a near circular orbit of radius about 43 AU. It has an estimated diameter of about 1110 km, roughly the size of Pluto's moon Charon and about one-tenth the size of Earth. Although smaller than Pluto, Quaoar is 100 million times greater in volume than all the asteroids combined. Quaoar is spherical and is a possible candidate for classification as a dwarf planet. . Like other KBOs, Quaoar's composition is thought to be mainly ice mixed with rock. The surface temperature of is estimated at −230 °C making it one of the coldest bodies in the Solar System. From the surface of this body, the distant Sun would appear as bright as Venus does from Earth. A satellite named Weywot, of about 100 km diameter was discovered orbiting Quaoar in February 2007, but little is known about it.

Another KBO is **2002 AW197**, which was discovered in January 2002 by a group of scientists led by Mike Brown. This object is about 750 km in diameter and orbits on an elliptical path between 41 AU and 53 AU from the Sun. This body takes about 325 years to orbit the Sun. Spectroscopic analysis shows a strong red colour but no water ice. It is a possible dwarf planet.

One of the strangest objects in the Kuiper belt is **Haumea** (previously 2003 EL61). This object is only half as large as Pluto but it is oval-shaped like an Australian or American football. It spins end-over-end every 4 h like a football that has been kicked. Haumea appears to be made almost entirely of rock, but with a glaze of ice over its surface. Astronomers have detected two tiny moons (Namaka and Hi'iaka) orbiting Haumea. By following the orbits of the moons astronomers have been able to determine that the mass of Haumea is about 32 % that of Pluto.

The odd shape of Haumea is thought to have been caused by a collision with another object early in its history. This collision knocked some of the original ice away from the surface of the body, leaving behind mostly rock. The impact also caused the body to spin rapidly and take the shape we see today. The small moons orbiting Haumea may have come from debris blown away during the collision.

In March 2007, astronomers announced they had have also found five other icy bodies in orbits similar to Haumea. Such families of objects are common in the asteroid belt, but this is the first group found in the Kuiper belt. All the fragments have a colour and proportion of water to ice similar to Haumea, and each fragment also has a surface that looks like it was once an internal region of the original object (Fig. 13.6).

Orcus (2004 DW) is a KBO discovered by Mike Brown and David Rabinowitz (USA) in February 2004. This body has an elliptical orbit around the Sun, similar to that of Pluto, and it takes about 245 years to orbit the Sun. At its closest approach it is 30 AU from the Sun, while its greatest distance is 48 AU. With a diameter

Fig. 13.6 The object Haumea (2003 EL61) is shaped like a football (Credit: NASA).

of between 760 km and 810 km Orcus is smaller than Quaoar. Its surface temperature is around −230 °C. Observations in infrared by the European Southern Observatory give results consistent with mixtures of water ice and carbon-based compounds. Orcus appears to have a neutral colour in comparison with the reds of other KBOs. In February 2007, a satellite, called Vanth, of size between 270 km and 380 km was discovered orbiting Orcus. Under the guidelines of the IAU naming conventions, objects with a similar size and orbit to that of Pluto are named after underworld deities; Orcus is a god of the dead in Roman mythology.

The KBO **Makemake** (previously 2005 FY9) is a large spherical object with a diameter of about 1430 km. Discovered in 2005 by the team led by Mike Brown, this object orbits the Sun once every 308 years in an eccentric and inclined orbit like Pluto's. Spectral analysis showed the surface to resemble that of Pluto but is redder. The infrared spectrum indicates the presence of methane, as observed on both Pluto and Eris. The body lacks a substantial atmosphere because of weak gravity. No satellites have yet to be detected around Makemake. In July 2008 Makemake was classified as a dwarf planet.

The Scattered Disc

The scattered disc is a sparsely populated region beyond the Kuiper belt, extending from 50 AU to as far as 100 AU and further. Objects in this region have highly eccentric orbits and are often wildly inclined to the orbital plane of the major planets. Two of the first scattered disc objects (SDO) to be recognised are 1995 TL8 (at 53 AU from the Sun) and 1996 TL66 (at 83 AU). Other objects detected include: 1999 TD10, 2002 XU93 and 2004 XR190 (at 58 AU). Some astronomers prefer to use the term 'scattered Kuiper belt objects' for objects in this region.

Many of the SDOs are doomed in the long term because, sooner or later, their highly eccentric orbits will carry them close to the giant planets to undergo more scattering. They may last a few million years or even 100 million years in their current orbits, but eventually Neptune will flip them nearer Uranus, Saturn or

Jupiter. These planets will, in turn, fling them outward, far beyond the Kuiper belt and into the Oort cloud, or out of the solar system entirely or closer to the Sun (where they will become comets).

One of the major scattered disc objects is **Eris** (2003 UB313 and previously known as Xena). The elliptical orbit of this body takes it to within 38 AU from the Sun and as far as 97 AU. The discovery of this object in 2003 prompted astronomers to decide on a definition of a planet. If Eris had been classed as a planet, there may have been as many as 15 planets in the solar system. In the end, the IAU decided on a definition that excluded Eris and also Pluto as major planets, instead classifying them as dwarf planets (Figs. 13.7 and 13.8).

Eris has a diameter of 2326 km, which makes it as large as Pluto. It is the largest object found in orbit around the Sun since the discovery of Neptune and its moon Triton in 1846. At times Eris is even more distant than Sedna (see below) and it takes more than twice as long to orbit the Sun as Pluto (560 years). In 2005, a

Fig. 13.7 Eris, the largest known scattered disc object, and its moon Dysnomia (Credit: NASA).

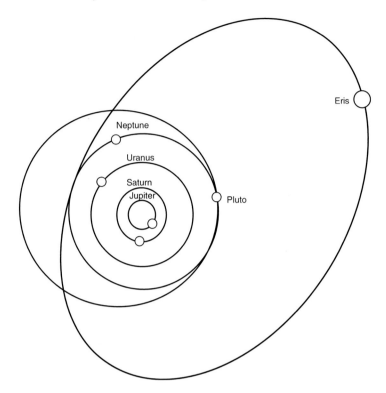

Fig. 13.8 Orbital path of Eris (2003UB313).

near infrared spectrograph on the Gemini Telescope in Hawaii, showed the surface of Eris to be mainly methane ice. Methane ice suggests a primitive surface unheated by the Sun since the solar system formed. If Eris ever had been close to the Sun, the methane ice would have been boiled off. Unlike the somewhat reddish Pluto and Triton, however, Eris appears almost grey. The interior of the dwarf planet is probably a mix of rock and ice, like Pluto. However, Eris is denser than Pluto.

The elliptical orbit of Eris is tilted at an angle of 44° to the orbital plane of the major planets. Eris has also been found to have a moon, named Dysnomia.

Eris is currently about three times Pluto's distance from the Sun, following an orbit that is about twice as eccentric and twice as steeply inclined to the plane of the solar system.

The Oort Cloud

The Oort cloud is an immense spherical cloud surrounding the solar system between 1000 AU and 100,000 AU (30 trillion km) from the Sun. This region contains billions of small icy objects probably left over from the formation of the solar system. Sometimes the orbit of one of these objects gets disturbed by other bodies, causing it to come streaking into the inner solar system as a long period comet (one with a period of around 2000 years). In contrast, short period comets take less than 200 years to orbit the Sun and they come from the Kuiper belt. The total mass of comets in the Oort cloud is estimated to be 40 times that of Earth.

One of the major Oort Cloud objects is **Sedna** (2003 VB12), which was discovered in November 2003 by a team led by Mike Brown at Palomar Observatory near San Diego, California, USA. The object was named after Sedna, the Inuit goddess of the sea, who was believed to live in the cold depths of the Arctic Ocean. Sedna has a highly elliptical orbit that is inclined at about $12°$ to the ecliptic. Its distance from the Sun varies between 76 AU and 937 AU, so it is best described as an inner Oort cloud object. Sedna will make its closest approach to the Sun (perihelion) about the year 2076 and will be furthest from the Sun (aphelion) in 8207. The shape of its orbit suggests it may have been captured by the Sun from another star passing by our solar system, or its orbit could be affected by another larger object further away in the Oort cloud.

Sedna is an odd body because no astronomers thought they would find an object like it in the empty space between the Kuiper belt and the Oort cloud. Spectroscopy has revealed that Sedna's surface composition is similar to that of some other Trans-Neptunian objects, being largely a mixture of water, methane and nitrogen ices with tholins. Sedna is the second reddish coloured object in the solar system after Mars. Its size is estimated to be about 995 km. Sedna has the longest orbital period of any known large object in the solar system, calculated at around 11,400 years. Recent estimates put its rotational period at about 10.3 h and its surface temperature at a very cold -250 °C. Sedna appears to have methane ice and water ice on its surface. The object's deep red spectral slope is indicative of high concentrations

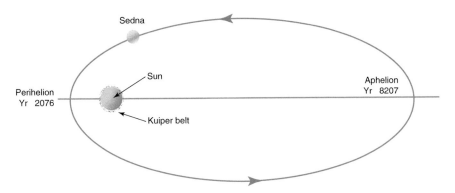

Fig. 13.9 Orbital path of Sedna.

of organic material on its surface, and its weak methane absorption bands indicate that methane on Sedna's surface is ancient, rather than freshly deposited. Models of internal heating via radioactive decay suggest that Sedna might be capable of supporting a subsurface ocean of liquid water. A search by the Hubble Space telescope has found no moons. Because Sedna has no known moons, determining its mass is currently impossible without sending a space probe.

Sedna is classified as a Scattered disc object, it currently is not a dwarf planet. Sedna has a Stern–Levison parameter estimated to be much less than 1, and therefore cannot be considered to have cleared the neighbourhood, even though no other objects have yet been discovered in its vicinity. To qualify as a dwarf planet, Sedna also must be shown to be in hydrostatic equilibrium (Fig. 13.9).

Another object, **2000 CR105**, of diameter 328 km, has a similar but less extreme orbit than Sedna: it has a perihelion of 44.3 AU, an aphelion of 416 AU, and an orbital period of 3491 years. It is considered a detached object.

In November 2014, astronomers announced the discovery of **2012 VP113**, an object half the size of Sedna in a 4268-year orbit similar to Sedna's and a perihelion within Sedna's range of roughly 80 AU. Its greatest distance from the Sun is around 450 AU. The surface of 2012 VP113 is believed to have a pink tinge, resulting from chemical changes produced by the effect of radiation on frozen water, methane, and carbon dioxide.

The similarity in the orbits found for Sedna, 2012 VP113, and a few other objects near the edge of the Kuiper belt suggests that an unknown massive perturbing body may be shepherding these objects into these similar orbital configurations.

Comets

Comets are small irregularly shaped objects thought to originate from the Kuiper belt, Scattered disc and Oort cloud. They are composed of a mixture of ice, dust and rock. When they are close enough to the Sun, they heat up, partially vaporise and develop a tail. The solar wind causes the tail to always point away from the Sun. Comets that are bright enough to be seen, cause much excitement when they appear in our night sky.

Comets are different to asteroids because they have an atmosphere surrounding their central nucleus and have a different origin. Most comets come from the outer regions of the solar system.

They usually orbit the Sun in a highly elliptical path and have a wide range of orbital periods ranging from several years to millions of years. Short-period comets generally originate in the Kuiper belt or the Scattered disc. Longer-period comets are thought to come from the Oort cloud.

Short period comets revolve around the Sun usually in the same direction as the planets and within 30° of the orbital plane of the planets. Long period comets are often found orbiting the sun at any inclination and may not orbit in the same direction as the planets.

Sometimes a long-period comet passes so close to a planet that the planet's gravitational force changes the comet's orbit, slowing it down and keeping it in the inner solar system. The comet then becomes a short period comet.

Some comets are found in circular orbits within the inner solar system.

Parts of a Comet

The solid core of a comet is known as the "nucleus". The nucleus is made up of rock, dust, and water ice with frozen gases such as carbon dioxide, carbon monoxide and ammonia. The surface of the nucleus is generally dry, dusty or rocky and appears dark. Comet nuclei with radii up to 30 km have been observed, but most are smaller. Because of their low mass, comet nuclei do not become spherical under their own gravity and therefore have irregular shapes. Not visible to the human eye is the hydrogen envelope, a sphere of tenuous gas surrounding the nucleus and measuring as much as 20 million km in diameter.

When a comet comes to within 3 AU from the Sun, heating causes streams of dust and gas to be released. The released material forms an atmosphere around the comet called the "coma". Gas and dust from the coma are always carried away from the Sun by the solar wind and radiation to form a "tail". Comets often display two distinct tails, depending on the comet's composition. One type of tail is made up of dust particles and appears yellow or white from reflected sunlight. The dust particles are very small (about 1 μm across) and move away from the nucleus into an arc that may be millions of kilometres long. The other type of tail is composed of ions (particularly carbon monoxide ions) and electrons—this tail is often straight, blue in colour, and extends about ten times farther than a dust tail. The ion tail is formed as a result of the ionisation by solar ultra-violet radiation of particles in the coma. Ionised particles are more affected by the Sun's magnetic field (see Figs. 13.10 and 13.11).

A comet is brightest and its tail is generally longest at about the time it passes perihelion (the closest point in its orbit to the Sun). As the comet moves away from the Sun it's tail, head and nucleus receive less solar radiation and gradually fade. The comet may be lost from view until its next return, which could be as short an interval as 3.3 years (as for Enke's Comet) or as long as 80,000 years (as for Comet Kohoutak) (Fig. 13.12).

Comets that survive passage around the Sun and return on a regular basis are said to be "periodic comets". However, periodic comets lose their mass reappearance after reappearance, and they

Fig. 13.10 Comet Hale-Bopp (C/1995 O1) as seen in 1997. The comet became one of the largest comets ever observed, with a nucleus measuring over 40 km in diameter. It is a long-period comet with an orbital period of 2537 years. Its greatest distance from the Sun is 370 AU. Notice the two parts to its tail (Credit: NASA).

eventually disappear. A typical comet loses between 1/60 and 1/100 of its mass with each pass of the Sun.

In October 1995, the comet Hale-Bopp was observed to eject some mass when it was still beyond Jupiter. Astronomers believe the ejection resulted from evaporation of surface ice with an assist from the comet's rapid rotation. Comets can also be destroyed when they come too close to a planet, a moon, or the Sun. A spectacular example of this was comet Shoemaker-Levy 9 which fragmented under the tidal force from Jupiter in 1992. Two years later, astronomers observed the pieces return and crash into Jupiter. The fact that the comet came apart in the first place, suggests that its nucleus was very weakly held together.

Other comets that have been observed to break up during their passage around the Sun, include Comet West and Ikeya-Seki.

If a comet passes across the Earth's orbit, then at that point there are likely to be meteor showers as Earth passes through the trail of debris. The Perseid meteor shower, for example, occurs

Fig. 13.11 Comet McNaught (C/2006 P1) was discovered by astronomer and comet-hunter Robert McNaught. This image of the comet was taken in January 2007 as both the comet and the Sun were setting over the Pacific Ocean. The comet has a very long period of 92,600 years and it reaches a maximum distance of 4100 AU from the Sun (Credit: European Southern Observatory).

every year between 9 August and 12 August, when Earth passes through the orbit of Comet Swift-Tuttle. Material from comet Halley is the source of the Orionid meteor shower in October.

Many comets and asteroids have collided with Earth in the past. Some scientists believe that comets that hit Earth about 4 billion years ago brought the vast quantities of water that exists in Earth's oceans. The detection of organic molecules in some comets has also caused some scientists to speculate that such comets may have brought life to Earth.

Probing Comets

The famous Halley's comet originates from the Kuiper belt and has a period of about 76 years. In 1986 five spacecraft were directed towards Halley as it approached Earth. Two of these missions were

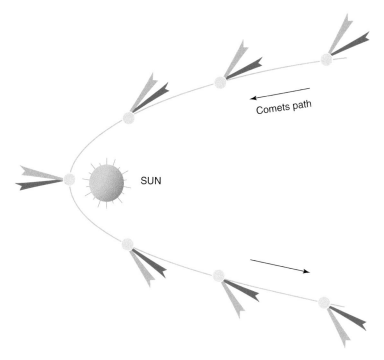

Fig. 13.12 The solar wind and radiation pressure from sunlight blow a comet's dust particles and ionised atoms away from the Sun. Consequently, a comet's tail always points away from the Sun.

launched by the Soviets (Vega 1 and 2), two by Japan (Susei and Sakigake), and one by the European Space Agency (Giotto). The probes found Halley's nucleus was irregular in shape (16 km long and 8 km wide). The dust-rich surface was darker than coal and contained small hills and craters. The comet's inner coma consists of a mixture of about 80 % water vapour, 10 % carbon monoxide, and 3.5 % carbon dioxide, and some complex organic compounds. Most of the particles in the tail are composed of a mixture of hydrogen, carbon, nitrogen, oxygen and silicates (see Table 13.4).

In 2001, the Deep Space 1 probe obtained high-resolution images of the surface of Comet Borrelly. The probe found the dark surface of this comet was hot and dry, with a temperature of between 26 °C and 71 °C.

In July 2005, an impactor from the Deep Impact spacecraft blasted a crater on Comet 9P/Tempel 1 to study its interior. The results of the mission suggested the majority of the comet's water

Table 13.4 List of comets visited by spacecraft

Comet name	Year discovered	Spacecraft	Year of visit	Closest approach (km)
Giacobini-Zinner	1900	ICE	1985	7800 km
Halley	Ancient times	Vega, Giotto	1986	8900 km
Grigg-Skjellerup	1902	Giotto	1992	200 km
Borrelly	1904	Deep Space 1	2001	2171 km
Wild 2	1978	Stardust	2004	240 km
Tempel 1	1867	Deep Impact	2005	Blasted a crater
Hartley 2	1986	EPOXI/DI	2010	700 km
Tempel 1	1867	Stardust	2011	181 km
Churyumov-Gerasimenko	1969	Rosetta	2014	In orbit, lander

ice was below the surface and that jets of vaporised water produced the coma of the comet. The probes spectrometer detected dust particles finer than human hair, and discovered the presence of silicates, carbonates, metal sulfides, amorphous carbon and polycyclic aromatic hydrocarbons (see Fig. 13.13).

Stardust was a 300 kg space probe launched by NASA on 7 February 1999. Its primary mission was to collect dust samples from the coma of comet Wild 2 and return them to Earth. The probe successfully completed its mission when it returned a canister containing the dust samples on 15 January 2006. The analysis showed that the coma/tail of comet Wild 2 contained a wide range of organic compounds (including the amino acid glycine) and crystalline grains that had been heated to a temperature around 1000 °C. In April 2011, scientists from the University of Arizona discovered iron and copper sulfide minerals that must have formed in the presence of water. The discovery shatters the existing paradigm that comets never get warm enough to melt their icy bulk.

The most exciting probe to reach a comet was the ESA's Rosetta probe in 2014. The three tonne spacecraft carried 11 experiments and a lander called Philae, which has another nine experiments. Rosetta obtained its power from two huge solar panels, each 14 m long. Since its launch in 2004, Rosetta had to make three gravity-assist flybys of Earth and one of Mars to help it on course to its rendezvous with comet 67P/Churyumov-Gerasimenko. In August 2014 Rosetta went into orbit around the 4 km wide comet. At this point Rosetta was 405 million km from Earth, about halfway between the orbits of Jupiter and Mars, and

Fig. 13.13 Image of comet 9P/Temple 1 taken 67 s after it was hit by Deep Impact's impactor in 2005 (Credit: NASA/JPL-Caltech/UMD).

moving at nearly 55,000 km/h. A lander called Philae was placed on the comet on 13 November 2014. The lander collected data with its suite of instruments, sniffing, hammering, drilling, and even listening to the comet. The data was sent back to Earth before the lander's battery ran out of power. Fortunately, Philae came out of hibernation on 13 June 2015 and was able to contact ESA scientists again. Philae found the comet is built up of fluffy layers of dust and ice with only half the density of water, and contains organic molecules necessary for life. The comets surface is riddled with unusual cylindrical pits that have walls covered in what scientists called "goosebumps".

Shaped like an enormous rubber duck, comet 67P has a large body that is joined to a smaller 'head' by a narrower 'neck' region. Scientists are unsure whether the comet gained its shape through uneven erosion, or as a result of two bodies fusing together after an ancient collision early in the solar system.

In early 2015, scientists announced that the deuterium to hydrogen ratio in water molecules surrounding the comet, was different to that of water molecules on Earth, suggesting comets like 67P were not the source of water on Earth.

Rosetta measured the comet's average temperature at $-70\,°C$, and found the comet was losing 11 kg of gas and dust a second as it approached the Sun. Scientists have not found any ice on the comet's surface. The dark exterior is covered with complex carbon-rich organic molecules.

The comet is in an elliptical 6.5 year orbit that takes it from beyond Jupiter at its furthest point to between the orbits of Mars and Earth at its closest to the Sun. Rosetta will accompany the comet for over a year as it swings around the Sun and back out toward Jupiter again (see Fig. 13.14).

A comet is generally named after its discoverer, the first person to see it (or the first two or three people if they independently find it at much the same time). Comets are also assigned letters and numbers. Many discoverers of comets are amateur astronomers who examine a particular area of the sky at night. Some comets are discovered not by the eye but by examination of photographs taken through telescopes. The European Space Agency's (ESA) SOHO spacecraft has been used to discover thousands of small comets that fly very close to the Sun. Many of them crash into the Sun or are pulled apart by its strong gravity.

Because new comets come from places in the solar system that are farthest from the Sun and thus coldest, they probably contain matter that is unchanged since the formation of the solar system 4.6 billion years ago. So the study of the constituents of comets is important for understanding the early stages of solar system formation.

Fig. 13.14 Comet 67P/Churyumov-Gerasimenko taken by the Rosetta probe on 6 August 2014 from a distance of 20 km. The image shows the comet's "head" at the left of the frame, which is casting shadow onto the 'neck' and 'body' to the right (Credit: ESA/Rosetta/MPS for ORIRIS Team/ NASA).

Meteoroids

The solar system is also strewn with rocky debris called meteoroids, which are smaller than asteroids. Most meteoroids are less than a few 100 m across. The ones larger than pebbles probably broke off when asteroids collided.

Meteorids are often pulled by gravity into the Earth's atmosphere. Air friction heats them up and their surface begins to vaporise. As the meteoroid penetrates further into the atmosphere, it leaves behind a trail of dusty gas, and we see it glowing as a meteor or shooting star. Some meteors even explode with a bright flash. Some even survive to hit the ground and are found by collectors as meteorites (see Table 13.5).

Table 13.5 Types of meteorites

Composition	Seen falling (%)	Finds (%)
Irons	6	66
Stony irons	2	8
Stones	92	26

Meteorites are grouped in three major classes according to their composition: iron, stony-iron, and stony meteorites. Rare stony meteorites called carbonaceous chondritis may be relatively unmodified material from the primitive solar nebula. These meteorites often contain carbon material and may have played a role in the origin of life on Earth.

Most meteorites that hit the Earth are stony in nature and are often referred to as simply stones.

Large meteorites that hit earth's surface produce a crater (similar to those we see on the Moon). The largest meteorite crater may be a depression over 400 km across deep under the Antarctic ice pack. This is comparable with the size of lunar craters. Another very large crater, in Hudson's Bay in Canada, is filled with water. Most meteorite craters on Earth are either disguised in such ways or have eroded away. A large crater that is obviously meteoritic in origin is the Barringer Crater in Arizona, USA. It is the result of what was perhaps the most recent large meteor to hit Earth, for it was formed only 25,000 years ago.

The Future

There is no doubt that more discoveries will be made in the Kuiper belt, the Oort cloud and Scattered disc. Any new objects discovered will be necessarily faint, cold and far away from Earth. It is also possible that the IAU will change their definition of what constitutes a planet in the future.

In August 2006, the IAU reclassified a number of objects in the solar system as 'dwarf planets'. Currently five dwarf planets are recognised by the IAU: Ceres, Pluto, Haumea, Makemake and Eris. Several other objects in both the Asteroid belt and the Kuiper belt are under consideration, with as many as 50 that could eventually qualify. Dwarf planets share similar characteristics to normal planets, but they are not dominant in their orbit around the Sun. The dwarf planets classified so far are members of larger populations. For example, Ceres is the largest body in the Asteroid belt, while Pluto is a large body in the Kuiper belt, and Eris is a member of the Scattered disc. Some objects might be considered to be dwarf planets, but their shape appears to deviate from hydrostatic equilibrium mainly because of massive impacts that occurred after they solidified. The definition of dwarf planet does not address this issue.

Under the 2008 IAU definition of planet, there are currently eight planets and five dwarf planets in the solar system. There has been some criticism of the new definition, and some astronomers have even stated that they will not use it. Part of the dispute centres around the idea that dwarf planets should be classified as normal planets. For now, the reclassification of Ceres, Pluto and Eris has attracted much media and public attention.

Further Information

https://solarsystem.nasa.gov/planets/profile.cfm (check out dwarf
    planets, comets, Kuiper belt, and Oort cloud)
www.space.com/planets/
www.pluto.jhuapl.edu
www.nasa.gov/mission_pages/newhorizons/

Glossary

Albedo A measure of the amount of light reflected from a planet, asteroid or satellite.

Aphelion The point in the elliptical orbit of a planet, comet or asteroid that is furthest from the Sun.

Apogee The point in the orbit of the Moon or artificial satellite at which it is furthest from Earth.

Apparent magnitude The visible brightness of a star or planet as seen from Earth.

Asteroid A small rocky and/or metallic object with a small size, often irregular in shape, orbiting the Sun in the asteroid Belt.

Asteroid belt A large group of small bodies orbiting the Sun in a band between the orbits of Mars and Jupiter.

Astronomical unit (AU) The mean distance between the Earth and the Sun, about 150 million km.

Atmosphere A layer of gases surrounding a planet or moon, held in place by gravity.

Aurora Curtains or arcs of light in the sky, usually in polar regions, caused by particles from the Sun interacting with Earth's or other planet's magnetic field.

Axial tilt The angle between a planet's axis of rotation and the vertical; equal to the angle between a planet's equator and its orbital plane.

Axis The imaginary line through the centre of a planet or star around which it rotates.

Big Bang An explosion that caused the birth of the universe about 13.7 billion years ago.

Black hole An object with gravity so strong that no light or other matter can escape it.

Cassini division Gap between Saturn's A and B rings.

Celestial equator An imaginary line encircling the sky midway between the celestial poles.

Celestial poles The imaginary points on the sky where Earth's rotation axis points if is extended indefinitely.

Celestial sphere The imaginary sphere surrounding the Earth, upon which the stars, galaxies, and other objects all appear to lie.

Chromosphere The layer of the Sun's atmosphere lying just above the photosphere (visible surface) and below the corona.

J. Wilkinson, *The Solar System in Close-Up*, Astronomers' Universe,
DOI 10.1007/978-3-319-27629-8,
© Springer International Publishing Switzerland 2016

Coma The diffuse, gaseous head of a comet.

Comet A small body composed of ice, rock and dust that orbits the Sun on an elliptical path.

Constellation One of 88 officially recognised patterns or groups of stars in the sky as seen from Earth.

Convection A heat-driven process that causes hotter, less dense, material in the Sun's interior to rise while cooler, denser material sinks.

Core The innermost region or centre of a planet or star.

Corona The tenuous outermost layer of the Sun's atmosphere, visible from Earth only during a solar eclipse.

Cosmology The branch of astronomy that deals with the origins, structure and space-time dynamics of the universe.

Crater A circular depression on a planet or moon caused by the impact of a meteor.

Crust The surface-layer of a terrestrial planet.

Dwarf planet Is a celestial body that is in orbit around the Sun, has sufficient mass for its self-gravity to overcome rigid body forces so that it assumes a hydrostatic equilibrium (nearly round) shape, has not cleared its neighbourhood around its orbit, and is not a satellite.

Earth Our home planet, one of the major planets of the Solar System.

Earthquake A sudden vibratory motion of the Earth's surface.

Eccentricity A measure of how elliptical an orbit is. A perfect circle has an eccentricity of zero, and the more stretched an ellipse becomes the closer its eccentricity approaches a value of 1 (a straight line).

Eclipse The total or partial disappearance of a celestial body in the shadow of another, such as a solar eclipse or lunar eclipse.

Ecliptic The apparent path of the Sun around the celestial sphere; also the plane of the orbit of the Earth around the Sun.

Electromagnetic radiation The name given to a range of radiations that travel at the speed of light. Includes infrared rays, ultraviolet rays, visible light, X-rays and gamma rays.

Ellipse The oval path, closed loop followed by a celestial body moving around another body under the influence of gravity.

Encke's division One of the narrow bands dividing Saturn's ring system. It is less prominent than the Cassini division.

Equator The imaginary line around the middle of a celestial body, half way between its two poles.

Escape velocity The minimum speed an object (such as a rocket) must attain in order to travel from the surface of a planet, moon or other body and into space.

Galaxy A huge group of stars, gas and dust held together by gravity and moving through space together.

Galilean moons The four largest moons of Jupiter discovered by Galileo Galilei in 1610. They are in order of distance from Jupiter, Io, Europa, Ganymede, and Callisto.

Gamma rays Form of electromagnetic radiation with short wavelength and high frequency.

Gas giant A large planet whose composition is dominated by hydrogen and helium. The gas giants planets in our solar system are: Jupiter, Saturn, Uranus and Neptune.

Gravity A force of attraction that exists between two masses.

Heliosphere A bubble blown into the interstellar medium by the pressure of the Sun's solar wind.

Ice Refers to solid states of water, methane, or ammonia which occur on planets or asteroids.

Inferior planet A planet that is closer to the Sun than Earth.

Infrared radiation Form of electromagnetic radiation with longer wavelength and lower frequency than the visible light region of the spectrum.

Ion An electrically charged atom or molecule, either positive or negative.

Ionosphere A layer of the Earth's atmosphere between 60 and 1000 km above the surface, where a percentage of the gases are ionised by solar radiation.

Kepler's laws Three laws discovered by Kepler that are used to describe the motion of objects in the solar system.

Kuiper belt A region of the solar system beyond Neptune (between 30 and 50 AU from the Sun), contains icy and rocky bodies similar to the asteroid belt.

Lava Molten rock flowing on the surface of a planet.

Lunar phase The appearance of the illuminated area of the Moon as seen from Earth.

Magnetic field A region of force surrounding a magnetic object.

Magnetosphere A region of space surrounding a planet or star that is dominated by the magnetic field of that body.

Mantle An inner region of a planet that lies between its crust and core.

Mare A plain of solidified lava on the surface of the Moon; appears darker than the surrounding area.

Mass The amount of material in a body; usually measured in grams or kilograms.

Meteor The bright streak of light that is seen when a rock or piece of space debris burns up as it enters Earth's atmosphere at high speed. Meteors that hit the Earth's surface are called meteorites.

Milky Way The galaxy of stars and gas clouds that our solar system belongs to, seen as a luminous band of stars across the night sky. It is a spiral galaxy.

Moon The only natural satellite of the Earth. The moon takes about 28 days to orbit the Earth once.

Nebula A cloud of gas or dust in the universe that may be illuminated by nearby stars.

Nuclear fusion A process whereby light atomic nuclei (such as hydrogen or helium) combine to produce heavier nuclei, with the release of energy; often called 'burning'. Occurs in stars but not planets.

Nucleus A collection of ices and dust that makes up the solid part of a comet.

Occultation The apparent disappearance of one celestial body behind another.

Oort cloud A sphere of icy bodies surrounding the outer solar system. Much further from the Sun than the Kuiper belt.

Orbit Path of one celestial body when moving around another.

Perigee The point in the orbit of the Moon or artificial satellite at which it is closest to Earth.

Perihelion The closest distance to the Sun in the elliptical orbit of a comet, asteroid or planet.

Period (of a planet) The time taken for a planet to orbit the Sun.

Photosphere The visible surface of the Sun or other star.

Planet A celestial body that is in orbit around the Sun, has sufficient mass for its self-gravity to overcome rigid body forces so that it assumes a hydrostatic equilibrium shape (becomes nearly round), and has cleared its neighbourhood around its orbit.

Planetesimal A small rocky or icy body, one of the small bodies that coalesced to form the planets.

Prominences Flame-like jets of gas thrown outwards from the Sun's chromosphere.

Protosun The part of the solar nebula that eventually developed into the Sun.

Radio telescope A telescope, often in the form of a dish-shaped receiver, designed to detect radio waves.

Radio waves Electromagnetic waves of low frequency and long wavelength.

Retrograde motion The apparent westward motion of a planet with respect to background stars.

Retrograde orbit An orbit of a satellite around a planet that is in the direction opposite to which the planet rotates.

Revolution The orbit of one body about another. One complete orbit is one revolution.

Rotation The spin of a planet, satellite or star on its axis.

Satellite Any small object (artificial or natural) orbiting a larger one, such as a moon orbiting a planet.

Scattered disc A distant region of our solar system, thinly populated by icy minor heavenly bodies.

Seasons The four divisions of the year of a planet whose axis of rotation is not perpendicular to the planet of its orbit. On Earth, the four seasons are: summer, autumn, winter and spring.

Shepherd satellite A satellite that constrains the extent of a planetary ring through gravitational interactions with the particles in the ring.

Sidereal time The orbital period of a planet or satellite as measured with respect to the stars.

Solar flare A sudden release of energy in or near the Sun's corona resulting in radiation being emitted into space.

Solar nebula The cloud of gas and dust from which the Sun and solar system formed.

Solar system The Sun, planets and their satellites, asteroids, comets, and related objects that orbit the Sun.

Solar wind A stream of charged particles or ions emitted by the Sun.

Space probe A spacecraft or artificial satellite used to explore other bodies (such as the planets or Moon) in the solar system. Such a craft contains instruments to record and send back data to scientists on Earth.

Space station A craft or vehicle that is in stable orbit around the Earth or other planet and is the temporary home of astronauts.

Star A self-luminous sphere of gas.

Sunspot A highly magnetic storm on the Sun's surface that is cooler than the surrounding area and so appears dark compared to the rest of the Sun.

Superior planet A planet that is more distant from the Sun than Earth is.

Supernova An exploding star, which briefly emits large amounts of light.

Tectonic forces Forces within a planet or moon that lead to the deformation of the crust of the body.

Terrestrial planet A planet whose composition is mainly rock (Mercury, Venus, Earth and Mars).

Transit The passage of one astronomical body in front of another, for example, when a planet passes in front of the Sun's disc as seen from Earth.

Weight The force of gravity acting on an object.

X-rays Electromagnetic radiation with short wavelength and high frequency (between ultraviolet and gamma rays).

Zodiac The name given to a group of twelve constellations that lie along the path followed by the Sun across the sky.

About the Author

John Wilkinson is a science educator with over 30 years experience in teaching science, physics and chemistry in secondary colleges and universities in Australia. He is author of over 100 science textbooks. He completed his Masters degree and PhD in science education at La Trobe University, Australia. Throughout his life he has been interested in Astronomy and operates his own observatory from his backyard. His main astronomical interests include the Moon, Sun and Solar System objects.

John is also author of "New Eyes on the Sun" and "The Moon in Close-up" both published by Springer.

J. Wilkinson, *The Solar System in Close-Up*, Astronomers' Universe,
DOI 10.1007/978-3-319-27629-8,
© Springer International Publishing Switzerland 2016

Index

J. Wilkinson, *The Solar System in Close-Up*, Astronomers' Universe,
DOI 10.1007/978-3-319-27629-8,
© Springer International Publishing Switzerland 2016

Printed in the United States
By Bookmasters